SpringerBriefs present concise summaries of cutting-edge research and practical applications across a wide spectrum of fields. Featuring compact volumes of 50 to 125 pages, the series covers a range of content from professional to academic.

Typical publications can be:

- A timely report of state-of-the art methods
- An introduction to or a manual for the application of mathematical or computer techniques
- A bridge between new research results, as published in journal articles
- A snapshot of a hot or emerging topic
- An in-depth case study
- A presentation of core concepts that students must understand in order to make independent contributions

SpringerBriefs are characterized by fast, global electronic dissemination, standard publishing contracts, standardized manuscript preparation and formatting guidelines, and expedited production schedules.

On the one hand, **SpringerBriefs in Applied Sciences and Technology** are devoted to the publication of fundamentals and applications within the different classical engineering disciplines as well as in interdisciplinary fields that recently emerged between these areas. On the other hand, as the boundary separating fundamental research and applied technology is more and more dissolving, this series is particularly open to trans-disciplinary topics between fundamental science and engineering.

Indexed by EI-Compendex, SCOPUS and Springerlink.

More information about this series at https://link.springer.com/bookseries/8884

Muhammad Bilal Tahir ·
Muhammad Shahid Rafique · Muhammad Sagir ·
Muhammad Faheem Malik

New Insights in Photocatalysis for Environmental Applications

 Springer

Muhammad Bilal Tahir
Department of Physics
Khwaja Fareed University of Engineering
and Information Technology
Rahim Yar Khan, Punjab, Pakistan

Muhammad Shahid Rafique
University of Engineering and Technology
Lahore, Pakistan

Muhammad Sagir
Department of Chemical Engineering
Khwaja Fareed University of Engineering
and Information Technology
Rahim Yar Khan, Pakistan

Muhammad Faheem Malik
University of Gujrat
Gujrat, Pakistan

ISSN 2191-530X   ISSN 2191-5318 (electronic)
SpringerBriefs in Applied Sciences and Technology
ISBN 978-981-19-2115-5   ISBN 978-981-19-2116-2 (eBook)
https://doi.org/10.1007/978-981-19-2116-2

This Springer imprint is published by the registered company Springer Nature Singapore Pte Ltd.
The registered company address is: 152 Beach Road, #21-01/04 Gateway East, Singapore 189721, Singapore

Contents

Chapter 1
A Fundamental Overview

Abstract Humans have always been fascinated by exploring new insights about the enigmatic tendencies of materials and their properties. However, after Feynman stated, "There is plenty of room at the bottom", nanoscience and nanotechnology have transformed the world into a new intriguing sector of science. Any material that has a size between 1 and 100 nm is considered to be a nanomaterial. However, nanomaterials can be tailored in various sizes and shapes by altering a number of different parameters. After the discovery of graphene in 2004, a new class of materials (2D) came into existence. By employing numerous synthesis strategies, two-dimensional nanorods and zero-dimensional quantum dots can also be prepared with great ease. Recently, nanomaterials have been introduced in the field of photocatalysis to effectively degrade pollutants, contaminants and dyes from water. Cocatalysts can be included into advanced composites and heterostructures to improve photocatalytic activity. Temperature and pH, in particular, have been proven to have an influence on the catalytic efficacy of any nanomaterial. This chapter provides a vision into the fundamentals of nanotechnology and the impact of nanomaterials in the field of photocatalysis.

1.1 Nanomaterials Chronology; From Bulk to 0D

Human history is so complex that some of the things that are being found and established yet need a definitive and accurate explanation. So, it is with nanomaterials and nanotechnology, which man has only recently become aware of, but whose roots may be traced back to the ancient Roman culture of the fourth century AD. The ancient antique piece named Lycurgus cup placed in a British museum is an excellent example of nanotechnology. It was made by using glass, and it is still the oldest form of dichroic glass. Dichroic glass can change its colour under specific conditions of light. Thus, the cup shows a variation in colour that it turns into green in the presence of direct light, and purple or red when the light shines through the glass. The lecture of Feynman entitled "There is plenty of room at the Bottom" in 1959 gave birth to the concept of nanotechnology. Feynman's theory led to different methods for

© The Author(s), under exclusive license to Springer Nature Singapore Pte Ltd. 2022
M. B. Tahir et al., *New Insights in Photocatalysis for Environmental Applications*,
SpringerBriefs in Applied Sciences and Technology,
https://doi.org/10.1007/978-981-19-2116-2_1

the fabrication of nanomaterial. These fabrication methods can be categorized into two different categories: in which one is top-down, and other is bottom-up. These categories are differentiated on the basis of speed, cost, and quality [1].

Drexler introduced a machine that can independently separate atoms and molecules and can produce stable nanostructures. Drexler, Peterson, and Pergamit published a book entitled "Unbounding the Future: The Nanotechnology Revolution" in which they introduced the term nanomedicine in 1991. The scientist who invented STM was awarded with a Nobel Prize in 1986. On the other side, the atomic force microscope and scanning probe microscope were invented in the same year. Carbon nanotubes offer a wide range of applications in nanotechnology due to their flexibility and strength. They have many applications in many various fields of science and technology, such as polymers, emitters, and energy storage devices to improve the different properties, among others. At the start of the twenty-first century, Feynman's idea of deploying matter was the most important feature of nanoscience and nanotechnologies [2].

Xu et al. discovered carbon dots with a size less than 10 nm while purifying single-wall carbon nanotubes in 2004. Carbon dots (C-dots) illuminated because of their unique qualities in the field of imaging and biosensors, such as low cost, low toxicity, and greater compatibility. C-dots have widespread applications in the field of electronics because they possess outstanding optical and electronic properties [3]. Nanoscience, on the other hand, entered in the field of computer science by size reduction of computers from a mass to efficient and well-designed laptops. Many advancements have come to see in the field of mobile phones and many other electronic devices. Nanotechnology is playing an essential role in the diagnosis and cure of human diseases. Many scientists believe that biotechnology is the most important application of nanoscience, and in the twenty-first century, nanotechnology has played a critical role in the pharmaceutical business.

The applications of molecular biology are related to nucleic acids. Paul Rothemund invented "scaffolded DNA origami" in 2006, which he constructed by enlarging well-self-assembled DNA. In 1982, for the very first time Nadriam Seeman introduced the concept of constructing a nucleic acid sequence. Scientists from various scientific fields, including physics, chemistry, biology, and computer science, come forward to safeguard the future of nanotechnology and to deal with expected future challenges. With the passage of time now, direct use of DNA to diagnose diseases by using nanotechnology is possible [4].

Nano-oncology and traditional drugs for chemotherapy have also been improved on a large scale to deal with the hazards of cancer in human. In this context, functional molecules such as nanodots, cytotoxic agents, nanoparticles, and antibodies have been presented. Various researches have been done to establish the employment of nanoparticles in the human body independently. The inclusion of therapeutic nanoparticles in the body can alter and regulate a variety of biological systems, including oxidative stress, metabolism, anticancer action, and autophagy. Thus, nanoscience has a significant impact on oncology, providing an efficient response rate for tumours and reducing toxicity, as toxicity is the major drawback of chemotherapy treatments [5].

Environmental applications are yet another fascinating aspect of nanotechnology, as it has the potential to improve energy resources while reducing pollutants. Nanoscience has provided sustainable efficient, effective, cheap, environment-friendly and reusable energy resources such as solar cells that give excellent performance at large scale with affordable prices. Nanotech-based solar cells also reduce organic chemicals and compounds that affect groundwater. Moreover, reduction of volatile organic compounds from the air has also become possible through these solar cells.

On the other hand, given how underdeveloped this discipline is, it is worth highlighting the essential need for improvement in computational field and nanomedicine. After a growing need for applications in the computational field, the field of nano-informatics was developed. The success of the nanotechnology would be the introduction of systems that can effectively learn and decode algorithms to assess future challenges using nanocarriers. Such designs can predict the prospective database, and these systems have yet only been applied for the prediction of cellular uptake and cytotoxicity of nanoparticles [6].

Nanoinformatics has so far been able to evaluate a substance's absorption, distribution, metabolism, excretion, and toxicity, through a process known as ADMET. Whereas, data mining, quantitative structure–property relationship (QSPR), and network analysis have also been predicted to be increased owing to their huge utilization in the field [7].

This field of nanoinformatics has not only provided an effective platform for developing and analytic functions using nanoparticles to overcome different obstacles, but it has also provided an effective platform for designing and analytic functions using nanoparticles to overcome various barriers. Moreover, drug-resistant tumours can be tricky to treat, so nano-informatics can assist in detecting such tumours. In this way, the procedure of chemotherapy can be improved. The minor side effects in the treatment of cancer are supported by the drugs delivered through nanoparticles based on hyperthermia [8].

1.1.1 Terminologies

To understand nanotechnology, a few terminologies such as nanoscale and nanoscience must be understood. The prefix nano is derived from the Greek term Nanos, which means "very short person". The International Standardization Organization (ISO) has provided the definition of nanoparticle (NP) "the object having all three dimensions in nanoscale which is 1–100 nm". However, nanomaterial (NM) is the object with one external dimension or internal structure and surface in nanoscale. The American National Standards Institute (ANSI) has established a set of proposed priorities known as the Nanotechnology Standards Panel (NSP). The following are the nanotechnology standardized groupings:

1. Terminology for materials composition and feature.
2. Terminology for nanoscience and nanotechnology.
3. Standard-test methods and analysis metrology.
4. Risk management and assessment, effects of toxicity and environmental impacts [9].

The ANSI-based nanotechnology panel has divided all these terminologies further into four groups that are given as follows:

1. Organizing all the terminotics according to composition and features.
2. A set of universal vocabulary to describe nanotechnology and nanoscience.
3. Analysis techniques.
4. Disastrous impacts and the risk management [10].

Thus, there is no single definition to describe nanomaterials internationally. Different definitions and terminologies for nanotechnology are given in the following aspects:

Nanoscale: The size range in this scale is 1–100 nm in all three dimensions. Only one dimension or all three dimensions can be in nanoscale for any materials.

All these dimension-based nanomaterials have four categories such as three-dimensional (3D), two-dimensional (2D), one-dimensional (1D), and zero-dimensional (0D) materials. Such materials can be in different sizes, compositions, states, and shapes depending upon their characteristics either bulk, powder or in solution form. A layered structure can also be found in some of the materials. Their vast applications in electronic and technical industries can also depend on the type and dimension of materials. These materials are categorized into the following types:

3D Materials: If all dimensions of a material are larger than nanosize (micro sized or greater), they are called 3D materials like a string or a ball [11].

2D Materials: If one of the dimension of any material is nanosized and the other two are not, such type of materials are called 2D materials like nanosheets [12].

1D Materials: When two dimensions of a material are nanosized, and one dimension of that material is in cm or mm, that type of material is known as 1D material like nanotube or nanowire [13].

0D Materials: A 0D substance, such as a quantum dot, has all three dimensions in the nanorange or that of nanosize. All these dimension-based morphological structures are visually described in Fig. 1.1 [14].

Nanoscience: Basically, this is the investigation of the matter in nanoscale. It also provides the difference between the properties of the material in bulk and in the nanostructure. The slight difference of properties in molecular, macromolecular and atomic level is also studied in this field.

Nanomaterials: If one external dimension of any material or any internal structure and surface structure is in nanoscale, then the material is said to be a nanomaterial.

Nano object: 1-, 2- and 3-dimensional objects at nanoscale can be confined in all three dimensions.

Nanoparticles: Because the longest and shortest axes for nano-objects with one, two, and three external dimensions differ in length at the nanoscale, so the terms nanorod and nanoplate could be used in place of nanoparticles [15].

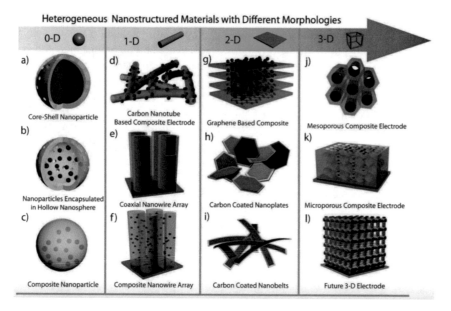

Heterogeneous Nanostructured Materials with Different Morphologies

0-D

1-D

2-D

3-D

a) Core-Shell Nanoparticle

b) Nanoparticles Encapsulated in Hollow Nanosphere

c) Composite Nanoparticle

d) Carbon Nanotube Based Composite Electrode

e) Coaxial Nanowire Array

f) Composite Nanowire Array

g) Graphene Based Composite

h) Carbon Coated Nanoplates

i) Carbon Coated Nanobelts

j) Mesoporous Composite Electrode

k) Microporous Composite Electrode

l) Future 3-D Electrode

Fig. 1.1 Structural complexity of nanomaterials based on dimension permission taken from Ref. [17] published by the permission of Royal Society of Chemistry

Nanofibers: Nano-objects have two identical dimensions, but the third is bigger than the others by an order of 100 nm or less. The particle having more than one dimensions at the nanoscale is nanofibers.

Nanostructure: Any structure with one or more dimensions measuring in the nanometer scale range that is 10^{-9} m (typically 1–100 nm).

Nanostructured material: The materials having structural parts in the range of 1–100 nm have multiple dimensions.

Nanocomposite: The composite consists of different phases in which one phase have more than one dimension [16].

1.2 History and Background of Photocatalysis

A photocatalytic reaction is one that takes place in the presence of light and involves a catalyst that absorbs light and participates in the process. In the late 1960s, the photoelectrolysis of water through TiO_2 was discovered, in which TiO_2 acted as an electrolyte electrode due to its unique property of oxidizing water into oxygen. The TiO_2 is a very stable material in the presence of an aqueous solution. In 1969, solar electrolysis was introduced, and this electrolysis was reported by many reputed journals. At the same time, prices of crude oil were suddenly raised and the availability of oil in the future became a serious issue; that's why that period is known as the oil crisis time. Then in 1970s, a number of publications were reported about this electrolysis process which evidenced a huge development in the field of photocatalysis. During the 1970s, the splitting of water was discovered by Fujishima and Honda. In the presence of a semiconductor catalyst, researchers reported the splitting of water into oxygen and hydrogen. In the late 1970s, energy crisis-related applications such as dye-sensitized solar cells were fuelled by using the photocatalysis process, which created significant interest in the field of photocatalysis. The scientists who gave the idea of photocatalysis in nanotechnology were awarded by Nobel Prize in 1996. At the start of the 2000s, many publications were reported to solve the issues in photocatalytic chemistry. When the light has less wavelength (415 nm or 3.0 eV), the band gap of TiO_2 electrode falls on the material's surface, then photocurrent from the external circuit flow from the platinum to TiO_2 electrode. The direction of current decides the oxidation reaction on TiO_2 electrode and reduction on the platinum electrode. This experiment displays that water can split up into oxygen and hydrogen in the presence of UV-visible light [18].

Thermodynamic equilibrium occurs at the interface in which electrode and electrolyte solution are in contact with each other and forms a thin surface charge region that can bend energy bands upward or downward in the case of n-type or p-type semiconductors, respectively. The width of the layer depends on the density of charge carriers and the dielectric constant of semiconductor material. If the electrode receives energy more than its band gap, then the generation of electron–hole pair and separation of charged layers occurs. In n-type semiconductors, the generated holes move to the interfacial region while electrons move towards the electrode and then electrically connected to the external circuit. In the case of p-type semiconductor, the generated holes will move towards the electrode while electrons are towards the interfacial region. The energy conduction band is larger than the evolution of hydrogen, and the electrons generated during the process flow towards the electrode. In case of other semiconductor materials, there should be a difference between the pH of anolyte and catholyte for the evolution of hydrogen [19]. The band gap of the semiconductor material should be near to the value of the energy that would be 1.35 eV. While for the water splitting, the required band gap should be 1.23 eV. Though many semiconductors were tested for the electrolysis of water, it wasn't successful with visible light due to their rusted electrolyte under irradiation. The generated holes oxidized themselves under the radiation. Dye sensitization of TiO_2 may employ a long wavelength of light, and photoelectrolysis of water through TiO_2 is theoretically feasible, but it is not practical. Photosensitizer dyes are most unstable below limitations. Regeneratable photoelectrochemical cells consist of one redox couple; further, the oxidized form is made at photo-anode and reduced at electrode [20]. Many toxic dyes were used as sensitizers in the past, but they were not stable neither efficient in those days, but late in the 1990s, this approach was highlighted by Gratzels et al. by utilizing porous electrode, which increases the efficiency from 10 to 11%.

The basic principle of photocatalysis is to put a semiconductor under visible or ultraviolet light; thus, light falls on semiconductor electrons transferred from the valence band to the conduction band. Thus, the holes are generated in the valence band. These photogenerated electrons and holes cause the production of free radicals that lead to the oxidation at valence and reduction in the conduction band. In this way, the targeted pollutant is degraded, and CO_2 is released. The main goal of improving this process is to increase the mobility of these light-induced couples and the rate of recombination. Lesser is the recombination rate of electrons and holes, greater will be the photocatalytic activity [21].

1.2.1 Photocatalysis for Environmental Pollution

Global pollution has been a major issue over the years due to industrialization. Its harmful consequences have impacted human health and posed a threat to all living species. Many organic and inorganic pollutants, including dyes, smokes, and metals, wastes have harmful effects and need to be resolved. Moreover, water is an essential element for the survival of all living organisms. However, among all the water resources on the planet, only a small portion is worth-drinking and clean. For the treatment of water and reduction of pollution, photocatalysis has served globally. Many semiconductors have been used to fix this problem; among them, the most famous are metal oxides, metal sulphides, oxynitrides, and carbon-based nanomaterials. Recently, after discovering two-dimensional (2D) materials, water treatment and pollution control have become more efficient through photocatalysis. The reported degradation of different hazardous dyes, including methylene blue, bisphenol and rhodamine blue, has become efficiently achievable [22].

1.2.2 Photocatalysis for Metals Reduction

Water pollution is limited to biomass, plastic, and dyes, but different metal particles, including zinc, titanium, chromium, ferric, and tin, have been present in water. Electroplating and other such industries have been causing this problem for a few years. Some forms of these metals are hazardous for human health when they enter to human body from aerosols. Hexavalent chromium can cause lung cancer and other skin disorders when it comes into contact with the human body. Because of all these issues, photocatalysis has provided a cheap and eco-friendly way to reduce these metals in their less harmful forms. Many semiconductors have been employed to reduce chromium, but ZnO has demonstrated to be particularly effective. Hexavalent chromium has been effectively reduced by ZnO to other forms such as trivalent chromium [22].

1.2.3 Photocatalysis for Energy Harvesting

The energy crisis has been one of the most alarming issues of the present age; thus, there is a colossal need to answer the issue. In order to overcome these crises, photocatalysis has provided an excellent solution. The applications of photocatalysis can provide production, storage and conversion of energy from one form to another. It has proved as an efficient, eco-friendly, affordable and accessible technique for the provision of solar energy. Solar cells, batteries, dye-sensitized solar cells, and supercapacitors have been improved by the use of different semiconductors and their nanocomposites.

Batteries based on 2D materials and nanocomposites have the ability for greater storage and better timing. Due to photocatalysis, many fuels can be converted from one form to another form. The significant feature of photocatalysis is the production of hydrogen energy to overcome the energy crisis [23].

1.3 Properties Affecting Catalytic Performance

The sanitization of dye molecules and the performance of photocatalysts are depended on the pH of the mixture. The pH of the solution can pedal the valence and conduction band positions, charge carriers and accumulation size. There are few materials that can be used as photocatalysts. It is determined by the material's morphology, and the size, shape, and structure of materials are important when selecting photocatalysts. Because the quality of a photocatalyst is to absorb the light, which is very important, so while selecting the photocatalyst, we have to check the absorption rate of the catalyst. The ratio of surface and volume should be very high so that the bombarding photon can strike on the maximum surface of the photocatalysts. If the volume-to-surface ratio is high, it can increase the area, which helps many photons strike on the surface. The photocatalytic activity depends on the band gap of the photocatalysts because the absorption rate of photons depends on the band gap of the semiconductor photocatalysts [24].

The placement of bands, including the conduction and valence bands, is the most critical factor in charge carrier redox calculations. The reduction reaction occurs at the most negative conduction band while the positive conduction band oxidation reaction occurs. Because a large specific area provides more active sites for light absorption and photocatalysis chemical reactions, a material with a large surface area will perform better. Adsorption of molecule is the main factor for the performance of photocatalysis reaction, so the large surface area allows the molecules to adsorb on the surface of the material. By changing the size, shape, dimension, and morphology of semiconductor, the surface area can be changed [25].

1.4 Impact of Incorporation of Other Materials

A certain quantity of photocatalysts could be used for the photocatalytic activity. As the amount of photocatalysts increases, it will increase the number of active sites, which leads to the best photocatalytic performance because there will be more space for the chemical reactions. There will be greater generation of OH radicals responsible for the breakdown of organic contaminants if there are more active sites. There will be more efficient degradation of the toxic pollutants when there are more OH radicals, but only a definite quantity of every photocatalyst can be utilized to degrade the pollutants. If a certain amount crosses the limit, it will thicken the solution and oppose the radiation to strike the surface of the specific semiconductor material or photocatalyst, which can cause a decrease in the efficiency of the process [26].

1.4.1 Effect of Doping, Composite, and Heterostructures of Semiconductors

The composites of the nanostructures are used for reducing the rate of combination of electron/hole pairs. Therefore, the separation of charge and efficiency of converting solar energy can be improved by the nanostructures. On the surface of TiO_2, metals act like electron sink, and this sink can create space between the surface of TiO_2 and the charge excited by the emitted light where electrons are placed on metal and holes are placed on TiO_2. The band structure with three-dimensional spacing of the excited electron and hole may be created by suitable placement. The rate of absorption and charge separation of semiconductors with a band gap less than TiO_2 are paired with TiO_2 to improve the rate of absorption and charge separation. For example, cadmium sulphide can be used as the photosensitizer for titanium dioxide to reduce CO_2. In the composite of CdS/TiO_2 photocatalysts, UV-visible light strikes on the surface of CdS and activates the active sites of the surface of CdS which further excite the electrons towards the conduction band of CDs. From the upper located conduction band of CdS, these excited electrons travel towards the lower located TiO_2 CB for the CO_2 reduction. By keeping the electrons on the TiO_2 and holes on CdS, the spatial space can be attained.

Solar energy conversion has achieved efficiency by using carbon materials of metallic character in composite-based photocatalysis. Most commonly, it has been observed that single-walled CNTs, i.e. carbon nanotubes have been used as electron sinks for enhanced conversion of solar energy. Graphene has been a fascinating material for photocatalysis since its exfoliation. In diverse morphologies, several graphene-based composite catalysts have been employed. Some thin-film nanocomposite of $TiO_2/graphene$ has also been fabricated that do not require any covalent modification to provide low graphene defect densities. On comparing single-phase TiO_2 with graphene-based TiO_2, the least efficiency has been observed in the case of TiO_2 than its counterpart. Twofold improved the efficiency for acetaldehyde oxidation for this nanocomposite. However, the same nanocomposite demonstrated a seven-fold improvement for CO_2 reduction compared with single-phase TiO_2 [27].

The previously mentioned graphene/TiO_2 nanocomposite technique lowers recombination, allowing electrons to be transferred to graphene. At the same time, the confinement of holes in TiO_2 makes the recombination slow. It has also been proved through many researchers that the higher mobility of graphene by electrical means can lead to enhanced photocatalytic activity of composite materials [28].

1.4.2   Effect of Metal Loading on Enhanced Activity

In case of reduction of CO_2, the efficiency can be enhanced by doping TiO_2 photocatalysts with the metals. Cu can effectively increase the efficiency of TiO_2 photocatalysts. The deposition of Pt nanoparticles can also improve the CO_2 reduction activity by using visible radiations. When plasmonic metals are used to cover a semiconductor, the rate of photocatalysis increases. This is due to the fact that metal may trap electrons and hence lower the rate of recombination [29].

1.5   Conclusion

Nanotechnology has been an area of remarkable creations that have benefited humanity in a variety of ways. In the past few decades, nanotechnology and nanoscience have made tremendous progress in revolutionizing the whole world. Miniaturizing from bulk to nanosheets, nanorods, nanotubes, and quantum dots has made scientists able to serve medical, electrons, computational, and agriculture. Different terminologies have been established in order to understand nanoscience globally. However, recently photocatalysis has revolutionized the field of nanoscience to resolve a number of environmental issues related to pollution, water treatment, metals reduction, and energy production.

However, a number of factors may have a remarkable impact on photocatalytic activity of any semiconductor. The worth mentioning properties are pH of catalyst, mobility of charge carriers, rate of recombination, temperature, and methods of synthesis. It has also been demonstrated that by including different catalysts into a composite or heterojunction, the overall photocatalytic activity may be significantly increased. However, elemental doping and metal loading can also aid the performance of any catalyst by reducing the recombination of electrons and holes.

References

1. Vijayalakshmi, R., Ramakrishnan, T., Srinivasan, S., & Kumari, B. N. (2020). Nanotechnology in periodontics: An overview. *Medico Legal Update, 20*(4), 2272–2277.
2. Sudha, P. N., Sangeetha, K., Vijayalakshmi, K., & Barhoum, A. (2018). Nanomaterials history, classification, unique properties, production and market. In *Emerging Applications of Nanoparticles and Architecture Nanostructures* (pp. 341–384). Elsevier.
3. Jeevanandam, J., Barhoum, A., Chan, Y. S., Dufresne, A., & Danquah, M. K. (2018). Review on nanoparticles and nanostructured materials: History, sources, toxicity and regulations. *Beilstein Journal of Nanotechnology, 9*(1), 1050–1074.
4. Castro, C. E., Kilchherr, F., Kim, D. N., Shiao, E. L., Wauer, T., Wortmann, P., Bathe, M., & Dietz, H. (2011). A primer to scaffolded DNA origami. *Nature Methods, 8*(3), 221.
5. Buttacavoli, M., Albanese, N. N., Di Cara, G., Alduina, R., Faleri, C., Gallo, M., Pizzolanti, G., Gallo, G., Feo, S., Baldi, F., & Cancemi, P. (2018). Anticancer activity of biogenerated silver nanoparticles: An integrated proteomic investigation. *Oncotarget, 9*(11), 9685.
6. Teleanu, D. M., Chircov, C., Grumezescu, A. M., & Teleanu, R. I. (2019). Neurotoxicity of nanomaterials: An up-to-date overview. *Nanomaterials, 9*(1), 96.
7. Norinder, U., & Bergström, C. A. (2006). Prediction of ADMET properties. *ChemMedChem: Chemistry Enabling Drug Discovery, 1*(9), 920–937.
8. Hua, X., Sun, Y., Chen, J., Wu, Y., Sha, J., Han, S., & Zhu, X. (2019). Circular RNAs in drug resistant tumors. *Biomedicine & Pharmacotherapy, 118*, 109233.
9. Hui, Y., Yi, X., Hou, F., Wibowo, D., Zhang, F., Zhao, D., Gao, H., & Zhao, C. X. (2019). Role of nanoparticle mechanical properties in cancer drug delivery. *ACS nano, 13*(7), 7410–7424.
10. Segev-Bar, M., Bachar, N., Wolf, Y., Ukrainsky, B., Sarraf, L., & Haick, H. (2017). Multi-parametric sensing platforms based on nanoparticles. *Advanced Materials Technologies, 2*(1), 1600206.
11. Zhou, X., Liu, B., Chen, Y., Guo, L., & Wei, G. (2020). Carbon nanofiber-based three-dimensional nanomaterials for energy and environmental applications. *Materials Advances, 1*(7), 2163–2181.
12. Xiong, J., Di, J., Xia, J., Zhu, W., & Li, H. (2018). Surface defect engineering in 2D nanomaterials for photocatalysis. *Advanced Functional Materials, 28*(39), 1801983.
13. Garnett, E., Mai, L., & Yang, P. (2019). Introduction: 1D nanomaterials/nanowires.
14. Wang, Z., Hu, T., Liang, R., & Wei, M. (2020). Application of zero-dimensional nanomaterials in biosensing. *Frontiers in Chemistry, 8*, 320.
15. Brown, K. A., Brittman, S., Maccaferri, N., Jariwala, D., & Celano, U. (2019). Machine learning in nanoscience: Big data at small scales. *Nano Letters, 20*(1), 2–10.
16. Alghoraibi, I., & Alomari, S. (2018). Different methods for nanofiber design and fabrication. In *Handbook of Nanofiber* (pp. 1–46).
17. Liu, R., Duay, J., & Lee, S. B. (2011). Heterogeneous nanostructured electrode materials for electrochemical energy storage. *Chemical Communications, 47*(5), 1384–1404.
18. Tahir, M. B., Kiran, H., & Iqbal, T. (2019). The detoxification of heavy metals from aqueous environment using nano-photocatalysis approach: A review. *Environmental Science and Pollution Research, 26*(11), 10515–10528.
19. Tahir, M. B., Malik, M. F., Ahmed, A., Nawaz, T., Ijaz, M., Min, H. S., Muhammad, S., & Siddeeg, S. M. (2020). Semiconductor based nanomaterials for harvesting green hydrogen energy under solar light irradiation. *International Journal of Environmental Analytical Chemistry*, 1–17.
20. Meng, A., Zhang, L., Cheng, B., & Yu, J. (2019). Dual co-catalysts in TiO_2 photocatalysis. *Advanced Materials, 31*(30), 1807660.
21. Weng, B., Qi, M. Y., Han, C., Tang, Z. R., & Xu, Y. J. (2019). Photocorrosion inhibition of semiconductor-based photocatalysts: Basic principle, current development, and future perspective. *ACS Catalysis, 9*(5), 4642–4687.

22. Khalid, N. R., Majid, A., Tahir, M. B., Niaz, N. A., & Khalid, S. (2017). Carbonaceous-TiO_2 nanomaterials for photocatalytic degradation of pollutants: A review. *Ceramics International, 43*(17), 14552–14571.
23. Hao, H., & Lang, X. (2019). Metal sulfide photocatalysis: Visible-light-induced organic transformations. *ChemCatChem, 11*(5), 1378–1393.
24. Liu, C., Dong, X., Hao, Y., Wang, X., Ma, H., & Zhang, X. (2017). Efficient photocatalytic dye degradation over Er-doped BiOBr hollow microspheres wrapped with graphene nanosheets: Enhanced solar energy harvesting and charge separation. *RSC Advances, 7*(36), 22415–22423.
25. Chen, J., Zhu, L., Xiang, Y., & Xia, D. (2020). Effect of calcination temperature on structural properties and catalytic performance of novel amorphous NiP/Hβ catalyst for n-Hexane isomerization. *Catalysts, 10*(7), 811.
26. Hu, Z., Mi, R., Yong, X., Liu, S., Li, D., Li, Y., & Zhang, T. (2019). Effect of crystal phase of MnO_2 with similar nanorod-shaped morphology on the catalytic performance of benzene combustion. *ChemistrySelect, 4*(2), 473–480.
27. Barrocas, B., Chiavassa, L. D., Oliveira, M. C., & Monteiro, O. C. (2020). Impact of Fe, Mn co-doping in titanate nanowires photocatalytic performance for emergent organic pollutants removal. *Chemosphere, 250*, 126240.
28. Rabhi, S., Belkacemi, H., Bououdina, M., Kerrami, A., Brahem, L. A., & Sakher, E. (2019). Effect of Ag doping of TiO_2 nanoparticles on anatase-rutile phase transformation and excellent photodegradation of amlodipine besylate. *Materials Letters, 236*, 640–643.
29. Sodagar, A., Akhoundi, M. S. A., Bahador, A., Jalali, Y. F., Behzadi, Z., Elhaminejad, F., & Mirhashemi, A. H. (2017). Effect of TiO_2 nanoparticles incorporation on antibacterial properties and shear bond strength of dental composite used in Orthodontics. *Dental Press Journal of Orthodontics, 22*(5), 67–74.

Chapter 2
Nanostructures Fabricated by Physical Techniques

Abstract Nanomaterials have received a lot of interest in recent years, and there has been huge research on different methods for the fabrication of nanomaterials. A few factors are worth noticing on the fabrication of any nanomaterials such as synthesis technique, availability of resources, cost of materials and techniques, efficient product, and time in which the required product is attained. The most commonly used methods are divided into four categories which include physical methods, chemical methods, electrochemical methods, and biological methods. All of the methods lack some of the above mentioned features. All approaches also include a set of techniques for the fabrication of nanomaterials such as sol–gel technique, coprecipitation, green synthesis, and laser ablation. Metal particles are mostly made using the sonochemical process, whereas nanosheets are made with chemical vapour deposition. However, for efficient production and reproduction of the required product, gamma irradiation is used.

2.1 Introduction

Nanoparticles are linked between the atomic or molecular structure and bulk material. Bulk materials have permanent physical properties, but in the case of nanoscale, material properties varied with the size of particles. The large surface area of nanoparticles is more prominent as compared to the bulk material. Solar cells built of nanoparticles are more efficient than thin-film solar cells. The nanoparticle-based solar cells can absorb more solar radiations by adjusting the morphology, including shape, size, and particle material. The absorption of solar light may be controlled by changing the size and form of nanoparticles. Because titanium dioxide is undetectable, it is utilised as a self-cleaning agent. Zinc oxides have extraordinary blocking properties, so they are used in the manufacturing of sunscreen lotions. Clay nanoparticles are tough and have a wide range of uses in plastics and polymer matrices. These clay nanoparticles convey their excellent hard property to polymers. For the designing of smart and advanced clothing, nanoparticles are also playing an important role in the field of textile fibres. On a nanometer scale, the quantum mechanical properties are

© The Author(s), under exclusive license to Springer Nature Singapore Pte Ltd. 2022   13
M. B. Tahir et al., *New Insights in Photocatalysis for Environmental Applications*,
SpringerBriefs in Applied Sciences and Technology,
https://doi.org/10.1007/978-981-19-2116-2_2

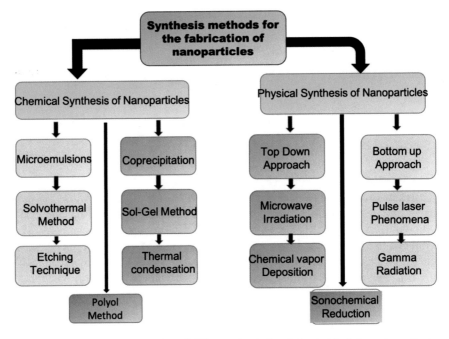

Fig. 2.1 Flow chart for a description of different chemical and physical techniques to synthesize nanoparticles

affected by the materials. At the nanoscale scale, essential features such as melting temperature, charge capacity, colour, form, and size may be altered without changing the chemical arrangement. Using these potentials will provide us more efficient and excellent performance of products that are only possible at nanoscale [1].

There are different methods for the fabrication of nanoparticles, such as microemulsion, microwave, sol–gel method, pulse laser ablation, spark discharge, template synthesis, biological synthesis, coprecipitation, hydrothermal synthesis, and inert gas condensation [2]. The following flow chart (Fig. 2.1) shows all of the procedures for production of nanoparticles, along with their explanations.

2.1.1   The Polyol Method

This is a chemical method used for the fabrication of nanoparticles, and a non-aqueous solvent (polyol) is used in this process, which also acts as a reducing agent. Polyol can minimize the oxidation of surface and accumulation. The shape, size, and texture of nanoparticles are adjustable. When the temperature is steadily increased while keeping the particle size constant, the polyol approach serves as a sol–gel process in the synthesis of oxides. Ethylene glycol has good reducing characteristics, as well as a high dielectric constant and a high boiling temperature. So, it is commonly used in the polyol method and also used as crosslinking reagents.

This method is also used for the fabrication of bimetallic alloys as well as core–shell nanoparticles. Yang et al. produced gold particles in the range of 100–300 nm by controlling the growth rate [3]. Xia et al. reported the fabrication of nanowires and cubes by adjusting the molar ratio of silver nitrate and PVP [4].

2.1.2 Microemulsions

The fine dispersion of a liquid into another liquid is known as an emulsion. The production of emulsion is possible through the solution of monomers/oils. Emulsions are divided on the basis of their droplet's size, including macro-, mini-, and microemulsions. For the preparation of inorganic nanoparticles, microemulsions are used on a large scale, as water and oil cannot form a mixture without any external source such as surfactant or emulsifier. So, they are phase separated, and to make them a single-phase or water–oil phase, input energy is required. This phase can be replaced by the water/water or oil/oil contacts. The surface tension between the oil and water could be high as 30–52 dynes/cm, and it can be minimized by surface-active molecule (surfactants). Surfactants have water-loving (hydrophilic) and oil-loving (lipophilic) groups. Surfactants can be used to identify the relationship and to stabilize the portions of oil and water by lowering the surface tension [5].

Maitra reported the fabrication of chitosan nanoparticles by using the microemulsion technique for the very first time. They prepared nanoparticles by adding a liquid solution of micellar droplets and cross-linked them by using glutaraldehyde. Surfactant mixed in N-hexane was also used in the preparation of chitosan and resolve the surface tension. After a non-stop stirring of all the material, nanoparticles were constructed [6].

Wongwailikhit and coworkers reported the preparation of iron oxide Fe_2O_3 in a micro-oil emulsion using H_2O (water). The size of the particles is determined by the amount of water in the emulsion, and in water–oil microemulsions, the water fraction may be raised to increase particle size [5].

2.1.3 Coprecipitation

The sudden occurrence of nucleation and growth is involved in the coprecipitation method. They exhibit the following characteristics:

I. Insoluble products are formed under supersaturation.
II. A great number of small particles could be formed during the nucleation process.

III.   The sudden change in size, shape, morphology, and property of the product is called a secondary process.
IV.   High saturation condition is a result of any chemical reaction.
V.   Aqueous solutions and metal oxides are constructed by combining aqueous and non-aqueous solutions, while metal chalcogenides are formed as a result of molecular precursor processes.

This approach has certain advantages as well as some problems. The advantages of this technology are precise and easy preparation, adjustable size and structure modification of the surface, lower temperature, efficient energy, and the absence of hazardous solvent. The inability to cope with uncharged particles, the inclusion of contaminants in the process, the time it takes, and the inability to reproduce are among the downsides. It doesn't work in case of different rate of participants [7].

2.1.4 Sol–Gel Method

One of the most beneficial techniques for the synthesis of the nanoparticle is supposed to be a sol–gel method. Many material properties such as surface structure, morphologies, and other physical properties can be altered through this technique. This approach may be used to create a variety of morphologies, including solid and ultra-thin films, glass fibre, and ceramics, during the synthesis of nanomaterials. This is done by providing low temperature in this chemical route.

Through this technique precursor is prepared, and after a series of steps, it is converted into a final product. The method includes drying, chemical composition and reactions, gel formation of precursors, and curing. Its easy handling, less time consumption, purity, control on doping concentration, and different morphological control make it a promising technique for the fabrication of nanoparticles [8].

2.1.5 Thermal Condensation

Condensation is the term used to define the change in the state of matter followed by some processes. When a liquid state of matter is changed into solid without passing through any further state, then that process is termed as deposition. However, nanoparticles can be fabricated through thermal condensation of different molecules; for example, if organic molecules are passed through thermal condensation, then residuals of carbon can be fabricated.

From the start, cyanamide wash has been used to make graphitic carbon nitride materials. At temperatures ranging from 230 to 335 °C, cyanamide is typically converted to dicyanamide, which is ultimately changed into melamine. If the temperature is further increased to 390 °C, then the melamine is rearranged; thus, the tri-s-triazine type is formed. In the end, g-C$_3$N$_4$ of polymeric form is obtained at 520 °C after further condensation [7].

2.1.6 Solvothermal

This type of synthesis technique involves chemical reactions based on solvents, especially organic solvents. An autoclave is used in the solvothermal method, and mostly, the coating of gold or platinum is used to provide extra cover and to avoid corrosion.

The process of preparation of material through the solvothermal method is based on crystallization through solution based on two steps. First step is crystal nucleation, and the second step is the subsequent growth of material.

Different morphologies and sizes can be obtained for any particular material by controlling the temperature of autoclave, pH of the solution, metal loading on the catalyst, and other compositions, including cocatalyst heterostructures. The solvothermal technique can provide an approachable and scalable route to synthesize the desired product.

Sometimes water, mainly distilled water, is used instead of other organic solvents to synthesize material because of the abundance. This process is termed as the hydrothermal method of synthesis [9].

2.1.7 Etching Technique

Thermal-oxidative etching is a technique to synthesize nanosheets from bulk material under air. A two-dimensional material, such as g-C$_3$N$_4$, has been created by gradually oxidative etching from the hydrogen strands that normally make up the fundamental unit of graphitic carbon nitride. However, on changing the time of the final process, product of desired thickness can be obtained. Desired results can be attained through this simple technique, whereas the whole process is temperature-dependent and is easily affordable [10].

2.2   Physical Synthesis of Nanomaterials

2.2.1   Bottom-Up

The fabrication of nanostructure from the bottom by using physical and chemical method is bottom-up technique. In this method, material is made by connecting atom by atom or molecule by molecule in nanosize ranging from 1 to 100 nm. Self-assembled atoms or molecules control the nanosize. Self-assembly is a bottom-up technique in which organized nanostructures are formed by physiochemical interactions. The technique in which single atom or molecule can be ordered one by one is known as a positional assembly [11].

2.2.2   Top-Down

Lithography and other sophisticated processes have been utilised to break down materials to the nanoscale in recent decades. The breaking of material into nanosize is called the top-down approach. The precision of engineer supports the microelectronics industry, and their performance can be enhanced by using the multiple improvements. Moreover, some specialized nanostructures based on cubic boron nitride and diamond are sometimes employed. Some sensors are also used to control the size, numerical analysis technique, and servo-drive technique. In lithography, the patterns on the surfaces can be drawn by using the light, ion, or electron beam to get the required material [12]. Different morphologies can be achieved by the use of different techniques (Fig. 2.2).

2.2.2.1   Microwave Irradiation

For the fabrication of inorganic–organic hybrid materials as well as for organic and inorganic materials, microwave irradiation method is used, due to its beneficial fabrication routes [48]. The oxidation of organic compounds by using the microwave method is reported. The mixture was prepared at room temperature and then irradiated by 3.67 GHz, 300 W microwave radiations. The reactions' progress was monitored by using chromatography (thin layer chromatography) and UV spectrophotometer at 352 nm.

Sahoo Biswa Mohan et al. combined phenylenediamine (1.08 g, 0.01 mol) and anthranilic acid (1.37 g, 0.01 mol) and further mixed in ethanol (15 ml). Then K_2CO_3 was added to a mixture, and the reaction mixture was placed in a microwave oven and refluxed at power (140 W) for 10 min. The TLC was used to observe the responses. After the distillation process was completed, the processed ethanol was filtered. Then the product was dried, filtered, and recrystallized from the ethanol [13].

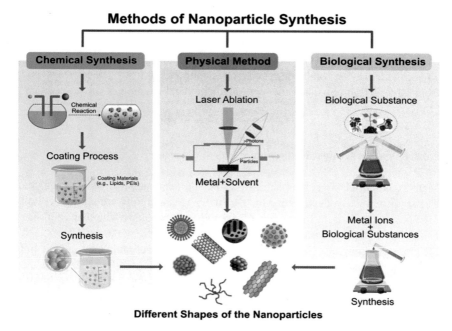

Fig. 2.2 Different methods of synthesis of nanoparticles of different morphologies [15]

2.2.2.2 Pulse Laser Phenomena

The construction of silver nanoparticles is done by using the pulsed laser method at very high production rate, which is approximately 3 gm/min. The reducing agent and silver nitrate solutions are mixed together in a blender with a solid disc that moves with the solution. A laser beam is used to create the spots on the disc's surface, while where the reaction of silver nitrate shows hot spots that are different from the centrifuge. Morphology can be controlled by the laser's energy and speed of disc [14].

2.2.2.3 Chemical Vapour Deposition

The chemical reaction is involved in the chemical vapour deposition (CVD). The CVD is used for the deposition of the thin films of different materials, e.g. gold and silver during the synthesis of the semiconductor materials. One or more precursors are used to get the desired deposition of films. At very high temperature, the precursor in the vapour form is inserted and deposited on the substrate. The absorbed molecules react with the existing molecules, and a new crystal is formed. The CVD method involves three-step process.

1. Transportation of reactants on the surface
2. Occurrence of chemical reaction on the surface

3. Removal of by-products from the surface. In gas, homogenous nucleation is
    formed, while in a substrate, heterogeneous nucleation occurs.

The chemical reaction occurred in the gas form can fabricate particle smaller than
1 μm, and it can control the morphology of nanoparticles as well. The size is ranged
from 10 to 100 nm by using the controlled chemical reaction of the gaseous phase
[16].

2.2.2.4 Sonochemical Reduction

Metal nanoparticles are fabricated by using the sonochemical reduction method.
The sonochemical reduction of MnO^{4-}, Au^{3+}, Au^+, and Pd^{2+} are reported with the
organic additives for the fabrication of controlled morphology. The shape and size of
nanoparticles can be controlled by controlling the reduction rate. Au^{3+} fabricated in
the presence of organic citric acid by controlling the reduction rate by sonochemical
reduction method [17].

Obreja et al. reported the fabrication of alcoholic reduction platinum by using
sonochemical method. Ethanol, propanol, and methanol were used as solvents and
reducing agents in the reduction of H_2PtC_{16}. The size of fabricated nanoparticle is
almost 3 nm [18].

2.2.2.5 Gamma Radiation

Metallic nanoparticles are synthesized by the gamma radiation method. This method
is very efficient because it has the property of reproducibility. The shape and size of
nanoparticle is controllable through the gamma rays. Non-toxic precursors are used
during the process, such as water, ethanol, and a smaller number of reagents. The
reaction steps are held at room temperature, which can minimize the by-products
and waste materials.

Radiolytic reduction is used to create highly scattered and mono-size metallic
clusters. For the fabrication of metallic clusters, radiolytic reduction is a potent
and efficient technique. When energetic gamma photons interact with the solution,
they ionize the solvent [19]. Abidi and Remita reported many reactions through this
method. After a series of reduction and oxidation process, water is produced. The
involved reaction step is given as:

The metallic nanoparticles are produced through the salt solution, and e^-_{aq} and
H$\bullet$ are the reducing agents of the process. Unluckily, the hydroxyl radicals OH$\bullet$
pannier the efficiency until hydroxyl hunters are used [20]. UV–vis spectroscopy is
used to find the plasmonic absorption band of MNP solutions of gold and silver. For
the trapping of MNP polymers, γ-rays' irradiation is used [21].

2.3 Conclusion

Nanomaterials have revolutionized the world in many ways; however, their fabrication and the choice of method is still a big question. A number of techniques can be used to fabricate nanomaterials; however, physical methods are considered more efficient to achieve the fine product. Chemical methods are utilized to produce nanomaterials on a large scale in addition to physical methods. From all of the chemical processes, the sol–gel and solvothermal procedures have gotten a lot of attention. Sol–gel technique involves easy steps, controlled temperature and a gel-like composition is made, and finally drying several hours results in nanoparticles of required material. The solvothermal method involves mixing and stirring of products and then placing in the autoclave at a certain temperature. Finally, washing, drying, and grinding of products give the required nanoparticles. Temperature, pressure, pH, and atmosphere in both cases can result in different morphological changes in the final product. CVD, laser ablation, sonochemical reduction, gamma irradiation, and microwave irradiation are the most well-known physical processes. Irradiation with gamma rays produces improved metallic particles with higher purity. Whereas, CVD is a three-step phenomena to fabricate nanoparticles on a substrate. All these processes have their own advantages and disadvantages. Thus, researchers have to find a method that is ecofriendly and consumes less time for better yield.

References

1. Landa, P., Dytrych, P., Prerostova, S., Petrova, S., Vankova, R., & Vanek, T. (2017). Transcriptomic response of Arabidopsis thaliana exposed to CuO nanoparticles, bulk material, and ionic copper. *Environmental Science & Technology, 51*(18), 10814–10824.
2. Reddy, K. R., Reddy, P. A., Reddy, C. V., Shetti, N. P., Babu, B., Ravindranadh, K., Shankar, M. V., Reddy, M. C., Soni, S., & Naveen, S. (2019). Functionalized magnetic nanoparticles/biopolymer hybrids: Synthesis methods, properties and biomedical applications. In *Methods in Microbiology* (Vol. 46, pp. 227–254). Academic Press.
3. Cele, T. (2020). Preparation of nanoparticles. In *Silver Nanoparticles-Health and Safety*. IntechOpen.
4. Wiley, B., Sun, Y., Mayers, B., & Xia, Y. (2005). Shaped-controlled synthesis of metal nanostructure: The case of silver. *Chemistry-A European Journal, 11*, 454–463.
5. Wongwailikhit, K., & Horwongsakul, S. (2011). The preparation of iron (III) oxide nanoparticles using W/O microemulsion. *Materials Letters, 65*, 2820–2822.
6. Maitra, A. N., et al. (1999). Process for the preparation of highly monodispersed hydrophilic polymeric nanoparticles of size less than 100 nm. US Patent 5874111.
7. Rane, A. V., Kanny, K., Abitha, V. K., & Thomas, S. (2018). Methods for synthesis of nanoparticles and fabrication of nanocomposites. In *Synthesis of Inorganic Nanomaterials* (pp. 121–139). Woodhead Publishing.
8. Amiri, M., Salavati-Niasari, M., & Akbari, A. (2017). A magnetic $CoFe_2O_4/SiO_2$ nanocomposite fabricated by the sol-gel method for electrocatalytic oxidation and determination of L-cysteine. *Microchimica Acta, 184*(3), 825–833.
9. Wang, C., Yang, K., Wei, X., Ding, S., Tian, F., & Li, F. (2018). One-pot solvothermal synthesis of carbon dots/Ag nanoparticles/TiO_2 nanocomposites with enhanced photocatalytic performance. *Ceramics International, 44*(18), 22481–22488.

10. Raagulan, K., Braveenth, R., Lee, L. R., Lee, J., Kim, B. M., Moon, J. J., Lee, S. B., & Chai, K. Y. (2019). Fabrication of flexible, lightweight, magnetic mushroom gills and coral-like MXene–carbon nanotube nanocomposites for EMI shielding application. *Nanomaterials, 9*(4), 519.
11. Yan, C., Yu, T., Ji, C., Kang, D. J., Wang, N., Sun, R., & Wong, C. P. (2019). Tailoring highly thermal conductive properties of Te/MoS2/Ag heterostructure nanocomposites using a bottom-up approach. *Advanced Electronic Materials, 5*(1), 1800548.
12. Smith, A. T., LaChance, A. M., Zeng, S., Liu, B., & Sun, L. (2019). Synthesis, properties, and applications of graphene oxide/reduced graphene oxide and their nanocomposites. *NanoMaterials Science, 1*(1), 31–47.
13. Fatimah, I. (2016). Green synthesis of silver nanoparticles using extract of Parkia speciosa Hassk pods assisted by microwave irradiation. *Journal of Advanced Research, 7*(6), 961–969.
14. Sadrolhosseini, A. R., Mahdi, M. A., Alizadeh, F., & Rashid, S. A. (2019). Laser ablation technique for synthesis of metal nanoparticle in liquid. IntechOpen.
15. Manawi, Y. M., Samara, A., Al-Ansari, T., & Atieh, M. A. (2018). A review of carbon nanomaterials' synthesis via the chemical vapor deposition (CVD) method. *Materials, 11*(5), 822.
16. Karuppasamy, L., Chen, C. Y., Anandan, S., & Wu, J. J. (2019). Sonochemical reduction method for synthesis of TiO2Pd nanocomposites and investigation of anode and cathode catalyst for methanol oxidation and oxygen reduction reaction in alkaline medium. *International Journal of Hydrogen Energy, 44*(58), 30705–30718.
17. Muthumariyappan, A., Rajaji, U., Chen, S. M., Chen, T. W., Li, Y. L., & Ramalingam, R. J. (2019). One-pot sonochemical synthesis of Bi2WO6 nanospheres with multilayer reduced graphene nanosheets modified electrode as rapid electrochemical sensing platform for high sensitive detection of oxidative stress biomarker in biological sample. *Ultrasonics Sonochemistry, 57*, 233–241.
18. Akman, F., Kaçal, M. R., Sayyed, M. I., & Karataş, H. A. (2019). Study of gamma radiation attenuation properties of some selected ternary alloys. *Journal of Alloys and Compounds, 782*, 315–322.
19. Klimov, D. I., Zezina, E. A., Lipik, V. C., Abramchuk, S. S., Yaroslavov, A. A., Feldman, V. I., Sybachin, A. V., Spiridonov, V. V., & Zezin, A. A. (2019). Radiation-induced preparation of metal nanostructures in coatings of interpolyelectrolyte complexes. *Radiation Physics and Chemistry, 162*, 23–30.
20. de Carvalho, L. M. J., Paranhos, B., de Jesus, E. F. O., & de Carvalho, J. L. V. (2020). Gamma Radiation Applied in Euterpe oleraceae Pulps. *European Journal of Nutrition & Food Safety*, 134–145.

Chapter 3
Nanomaterials for Safe and Sustainable Environment: Realm of Wonders

Abstract Nanotechnology is a very important and safest technology nowadays. From the past few years, nanotechnology has attracted a lot of attention because of researchers owing to its unique properties and versatile applications. Basically, nanomaterials have unique properties like small size (1–100 nm), large surface area, excellent mechanical, chemical, optical, electrical, and magnetic properties as compared to bulk materials. Nanomaterials play an essential role in various fields like environmental applications, energy applications, agriculture, and food storage. It is also critical in biology or medical industries and has a lot of other applications. This chapter begins with a discussion of nanotechnology, critical comparison of nanomaterials and their uses, followed by an analysis of nanomaterial classification based on dimensions, and finally a review of nanotechnology's future prospects.

3.1 Introduction

Nanotechnology is an emerging technology in the twenty-first century [1]. It is growing day by day due to its excellent properties and applications. In nanotechnology, the products are made on the nanoscale, so these are lighter in weight, easy to handle, possess high chemical stability, high magnetic stability, among others [2]. That's why nanotechnology is used in energy, environmental, food preservation, medical, and in many different industrial applications. One of the main reason that nanomaterials are used in different fields is due to the fact that they can be synthesized at nanoscale 1–100 nm. These materials have variable properties such as chemical, physical than bulk materials [3]. Nanomaterials are used as a sensor, catalyst, as well as in environmental applications owing to it nature of non-toxicity, environmentally friendly, and improved efficiency [4]. These materials are also used for the treatment of wastewater, for cleaning the air, and soil. By using nanomaterials, heavy metals, bacteria, dyes, and other pollutants can be easily removed from wastewater. The metal-based nanomaterials are very efficient for such environmental applications [5].

3.2 Insight in "The Room at Bottom" and Its Realm

The seed of nanotechnology is firstly grown in the talk of Richard Feynman in 1959, where he says, "There's plenty of room at the bottom" [6]. This means there is the possibility of directly synthesized material on a very small scale. In 1974, Norio Taniguchi discovered the word nanotechnology [7]. After that, nanotechnology has been grown rapidly day by day. Nano means small at 10^{-9}, this is very hard to imagine to synthesize the materials at that small level, but nanotechnology finds its way and emerged fast [8]. Nanomaterials have a nanoscale size, which is the key distinction that makes them more efficient [9].

In the nanomaterials or nanoparticles, the more atoms are present on the surface area as compared to the inner side; there is a simple example for this if the size of nanoparticle is 40 nm; there is 3.5% atom on the surface if the size is 10 nm; there is 20% atom on the surface; and if the size is 2 nm, there is 60% of atom on the surface [10]. Two main factors that can make nanomaterials different from bulk are the size of nanomaterials and the quantum effect [11]. Many additional properties of nanoparticles, such as magnetic properties, electric properties, optical properties, mechanical properties, and shape, differ significantly from those of bulk materials [12].

There are two main techniques for the synthesis of nanomaterials; "top-down" [13] and "bottom-up" [14]. In the "top-down" technique, the large compound or molecule is broken down into small materials. In the "bottom-up" technique, the small particles come together and form a complex or big structure. The nanomaterials are classified into four different categories, e.g. zero-dimensional, one-dimensional, two-dimensional, and three-dimensional [15].

3.2.1 3D Nanomaterials for Environmental Applications

The nanomaterials can be classified on the basis of many different parameters like size, shape, chemical properties, strength, morphology, magnetic properties, electric properties, optical properties, among others. The three-dimensional nanomaterials are those nanomaterials that are not limited to the nanoscale in any dimension [16]. In three-dimensional nanomaterials, all three directions are in macroscale. These materials have a large thickness, and having length which is above a few nanometers [17]. These materials carry bulk powder, large nanotubes, large nanowires, and multinanolayers.

These materials can also be synthesized by the "bottom-up" technique. In the three-dimensional graphene [18], carbon fibres are the most important material for environmental applications because of their unique properties. The carbon fibres can be synthesized by various techniques like the hydrothermal method, chemical vapour method, etc. [19]. These can be used on a large scale for environmental applications like air purification, water purification, and radiation absorption. As with carbon fibre three-dimensional nanomaterials, the graphene-based three-dimensional nanomaterials are also used for the treatment of these environmental problems [20]. These materials have many advantages such as non-toxic, environmentally friendly, cost-effective, high mechanical properties, lighter in weight, and high surface area.

3.2.2 2D Nanomaterials as the Star of Photocatalytic Applications

In the two-dimensional nanomaterials, the two dimensions are outside the nanoscale. In two-dimensional materials, only one direction is in nanoscale [21]. Nanomaterials with two dimensions have many unique properties, that's why they are widely used in a variety of applications such as environmental, energy, medicinal, and many more [22]. The examples of this class of nanomaterials are nanofilms, nanocoating, nanolayers, and 2D graphene [23]. The thickness of two-dimensional nanomaterial is at least one atomic layer. "Surface area" is the basic difference between the bulk and two-dimensional materials. The surface area of nanomaterials is high than the bulk. As a result of the large surface area, there are more atoms on the surface than on the inside, and these surface atoms serve a specific role. Nanomaterials are more efficient than bulk materials because of this. The applications or uses of two-dimensional materials are increasing day by day [24]. The two-dimensional hexagonal boron nitride, 2D graphene, and metal dichalcogenides are the most efficient 2D materials that have attracted huge attention as a catalyst, in electronics devices, in optic electronic devices, energy applications, environmental applications, and as a sensor, etc. owing to their excellent properties. As a catalyst, they show excellent properties like high surface area, excellent mechanical properties, high chemical stability, including others [22, 25]. Graphene as a semiconductor has 0 band gap and has a high charge carrier rate at room temperature.

These materials can be synthesized by various methods like chemical method, liquid exfoliation, and plasma-enhanced chemical vapour method [26]. There are many advantages of two-dimensional materials like environmentally friendly, cost-effective, lighter in weight, high surface area, and non-toxicity.

3.2.3 1D Nanomaterials for the Survival of Environment

In these materials, two dimensions are in nanoscale [27]. The one-dimensional nanomaterials have their own unique and excellent properties. In these materials, the surface area is larger than 3D and 2D nanomaterials. Again, if the surface area is larger, there are more atoms on the surface; consequently, they show more efficient properties as compared to 3D and 2D nanomaterials [28]. The 1D nanomaterials are used as a catalyst in photocatalysis, as a sensor in optical devices, as well as in environmental applications to purify water and air, and in energy applications [29]. The one-dimensional nanofibres, nanotubes, nanowires, and nanorods have attracted huge attention because of their versatile applications.

These materials can be synthesized by various methods such as chemical, hydrothermal method and reducing technique, etc. one-dimensional nanomaterials such as metals, oxides, and nitrides have great characteristics and are frequently employed in many innovative approaches [28]. In present times, one-dimensional carbon also attracts a lot of attention because it also indicates remarkable properties in all fields. As a semiconductor, ZnO is a very good photocatalyst and shows excellent properties as large surface area and mechanical properties, etc. [30]. The one-dimensional nanomaterials have picked a lot of applications in environmental science due to their advantages of cost-effective, non-toxic, environmentally friendly, large surface area, and excellent mechanical properties [31].

Nowadays, the environment is highly polluted by a variety of toxic contaminants, and our air is contaminated by poisonous gases that directly affect humans. The water and soil have been contaminated by several pollutants like heavy metals, dyes, and organic pollutants. All of these hazardous pollutants have the potential to harm people and other living species, as well as induce life-threatening illnesses such as cancer and chronic obstructive pulmonary disease. As a result, we utilize non-toxic and environmentally friendly nanomaterials that are appropriate for environmental applications since there is a need for time to treat the environment [32].

3.2.4 Perspective of 0D Nanomaterials in Photocatalysis

In the zero-dimensional nanomaterials, all dimensions are in nanoscale. The zero-dimensional nanomaterials have a very large surface area as compared to 3D, 2D, 1D nanomaterials; these materials are ultrathin in size [33]. The zero-dimensional nanomaterials are spherical-shaped nanomaterials with a diameter less than 100 nm. The materials include in zero dimensions are quantum dots, fullerenes, carbon quantum dots, graphene quantum dots, magnetic nanomaterials, polymer dots, etc. [34]. They offer great mechanical, optical, magnetic, and chemical properties, which is why they're employed as sensors in biological applications, and catalysts in energy applications. among others [35].

In photocatalysis, these nanomaterials are used as an excellent catalyst; they improve the process of photocatalysis and easily control the morphology. The carbon nitride nanotubes (CNNs) show excellent properties as a cocatalyst. When the CNNs combine with the photocatalyst, they show excellent results and also increase its photocatalytic activity. As a result, when zero-dimensional nanomaterials as a host are paired with CNNs, they provide exceptional results due to their large surface area, short charge transfer length, and easy control of material size, among other factors. Carbon dots behave like electrons, increasing the electron–hole pair's uncoupling ratio and noble metal atoms' light absorption [36].

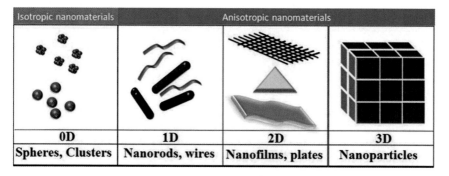

Fig. 3.1 Classification of nanomaterials on the base of dimensions (permission taken from Ref. [37] published after permission from Elsevier)

There are also many advantages of zero-dimensional nanomaterials, like they possess very large surface area. They are very thin in size, environmentally friendly, lighter in weight, cost-effective, and have excellent chemical, magnetic, mechanical, and optical properties. The novel technology combines the two-dimensional nanomaterials with zero-dimensional nanomaterials for producing more efficient nanomaterials used in various applications (see Fig. 3.1) [37].

3.3 Future Perspective in the Field of Nanotechnology

From the last few years, nanotechnology plays a vital role in various fields like electronics, energy applications, environmental applications, biological applications, and many more different applications. Nanotechnology, in particular, might play a critical role in the future, since this sector continues to grow in importance. To take forward this technology, we need to make the material on a very small scale means on the nanoscale, and this is an actual challenge for us. In the medical field, nanotechnology is already playing an important role in nanomaterials used as a sensor, imaging, drug delivery, etc. In the future, it is possible to make a sensor that shall work as a doctor inside our human body. These types of sensors will be injected into the human body for collecting information in a very short time. These sensors supposed to detect the affected area and will deliver medicines, without affecting the human body [38].

Nanotechnology is employed in cancer therapy; this technology makes cancer treatment easier and less unpleasant than in the past. This technology is little bit expensive nowadays, but in future it will be cheaper as research in this area is extensive. The nanomaterials are also used as detective sensors in data storage devices. In the future, it is possible to make a sensor that will detect big data and alert about the problem. It will also be possible to make a memory storage device that will store this large amount of data. Nanotechnology plays a significant role in converting solar energy into electricity. In the future, it might be possible to make a very efficient solar panel that will convert more solar energy into electricity, and it is also possible to make batteries that will store this energy for electric cars. Nanotubes and nanowires are the most efficient and practical materials for this purpose. These materials have a wide surface area, which allows them to store more energy and ensure that devices function properly [39].

For water purification, nanotechnology also plays a key role, and the nanomaterials are used as good purification agent for the treatment of water. In the future, it will also be possible to make more efficient nanomaterials that work more efficiently for the treatment of water [40]. The treatment of water is very important because it is an essential element for the survival of life. Nanotechnology also plays an essential role in agriculture and food storage. In food storage, nanomaterials use for a coating on the food, by the coating of nanomaterials, the environment cannot affect food so, that's why the food cannot be damaged and can be easily stored. In the future, it is possible to make a more efficient nanomaterial for a coating that can increase more lifetime of food. It can also be more cost-effective. So, in short, nanotechnology plays an essential role in the future because of its excellent properties.

3.4 Conclusion

This chapter gives detailed information about the use of nanotechnology in environmental applications. This technology has many positive effects on humans. Nanotechnology may be utilized to distribute drugs and can potentially be used to cure cancer. By using this technology, the treatment process is less painful and has less side effects. The nanomaterials take a lot of attention than bulk materials because of their excellent properties like large surface area, small size (1–100 nm), excellent mechanical, chemical, optical, magnetic, and electric properties. This technology created a plethora of new approaches in a variety of disciplines. Nanomaterials may be synthesized using two different techniques: "top-down" and "bottom-up." These two main techniques further divided into two subcategories. In a nutshell, nanotechnology has taken a lot of attention in the last few years, and this technology has a very bright future in all the field of science and technology.

References

1. Poole Jr, C. P., & Owens, F. J. (2003). *Introduction to Nanotechnology*. Wiley.
2. Bhushan, B. (2017). *Springer Handbook of Nanotechnology*. Springer.
3. Roco, M. C. (2011). *The Long View of Nanotechnology Development: The National Nanotechnology Initiative at 10 Years*. Springer.
4. Wiesner, M., & Bottero, J. -Y. (2007). *Environmental Nanotechnology*. McGraw-Hill Professional Publishing.
5. Ramsden, J. (2016). *Nanotechnology: An Introduction*. William Andrew.

6. Feynman, R. P. (1960). There's plenty of room at the bottom. California Institute of Technology, Engineering and Science magazine.
7. Roukes, M. (2001). Plenty of room, indeed. *Scientific American, 285*(3), 48–57.
8. Billinge, S. J. (2008). Nanoscale structural order from the atomic pair distribution function (PDF): There's plenty of room in the middle. *Journal of Solid State Chemistry, 181*(7), 1695–1700.
9. Roduner, E. (2006). Size matters: Why nanomaterials are different. *Chemical Society Reviews, 35*(7), 583–592.
10. Cao, G. (2004). Nanostructures & Nanomaterials: Synthesis, Properties & Applications. Imperial college press.
11. Ungár, T., Schafler, E., & Gubicza, J. (2009). Microstructure of bulk nanomaterials determined by X-ray line-profile analysis. *Bulk Nanostructured Materials*, 361–383.
12. Baimova, Y. A., Murzaev, R., & Dmitriev, S. (2014). Mechanical properties of bulk carbon nanomaterials. *Physics of the Solid State, 56*(10), 2010–2016.
13. Fu, X., et al. (2018). Top-down fabrication of shape-controlled, monodisperse nanoparticles for biomedical applications. *Advanced Drug Delivery Reviews, 132*, 169–187.
14. de Oliveira, P. F., et al. (2020). Challenges and opportunities in the bottom-up mechanochemical synthesis of noble metal nanoparticles. *Journal of Materials Chemistry A, 8*(32), 16114–16141.
15. Amarnath, C. A., et al. (2013). Nanohybridization of low-dimensional nanomaterials: Synthesis, classification, and application. *Critical Reviews in Solid-State and Materials Sciences, 38*(1), 1–56.
16. Van Gough, D., Juhl, A. T., & Braun, P. V. (2009). Programming structure into 3D nanomaterials. *Materials Today, 12*(6), 28–35.
17. Ersen, O., et al. (2015). Exploring nanomaterials with 3D electron microscopy. *Materials Today, 18*(7), 395–408.
18. He, S., & Chen, W. (2015). 3D graphene nanomaterials for binder-free supercapacitors: Scientific design for enhanced performance. *Nanoscale, 7*(16), 6957–6990.
19. Liu, P., et al. (2017). Recent progress in the applications of vanadium-based oxides on energy storage: From low-dimensional nanomaterials synthesis to 3D micro/nano-structures and free-standing electrodes fabrication. *Advanced Energy Materials, 7*(23), 1700547.
20. Zhou, X., et al. (2020). Carbon nanofiber-based three-dimensional nanomaterials for energy and environmental applications. *Materials Advances, 1*(7), 2163–2181.
21. Zhang, H. (2015). Ultrathin two-dimensional nanomaterials. *ACS Nano, 9*(10), 9451–9469.
22. Khan, A. H., et al. (2017). Two-dimensional (2D) nanomaterials towards electrochemical nanoarchitectonics in energy-related applications. *Bulletin of the Chemical Society of Japan, 90*(6), 627–648.
23. Li, M., Luo, Z., & Zhao, Y. (2018). Recent advancements in 2D nanomaterials for cancer therapy. *Science China Chemistry, 61*(10), 1214–1226.
24. Zhang, H., Chhowalla, M., & Liu, Z. (2018). 2D nanomaterials: Graphene and transition metal dichalcogenides. *Chemical Society Reviews, 47*(9), 3015–3017.
25. Ganguly, P., et al. (2019). 2D nanomaterials for photocatalytic hydrogen production. *ACS Energy Letters, 4*(7), 1687–1709.
26. Fojtů, M., Teo, W. Z., & Pumera, M. (2017). Environmental impact and potential health risks of 2D nanomaterials. *Environmental Science: Nano, 4*(8), 1617–1633.
27. Garnett, E., Mai, L., & Yang, P. (2019). *Introduction: 1D Nanomaterials/Nanowires*. ACS Publications.
28. Jin, T., et al. (2018). 1D nanomaterials: Design, synthesis, and applications in sodium-ion batteries. *Small (Weinheim an der Bergstrasse, Germany), 14*(2), 1703086.
29. Huang, L. B., Xu, W., & Hao, J. (2017). Energy device applications of synthesized 1D polymer nanomaterials. *Small (Weinheim an der Bergstrasse, Germany), 13*(43), 1701820.
30. Zhong, Y., et al. (2020). Interface engineering of heterojunction photocatalysts based on 1D nanomaterials. *Catalysis Science & Technology*.
31. He, Z., et al., 1D/2D Heterostructured photocatalysts: From design and unique properties to their environmental applications. *Small*, 2005051.

32. Yang, Y., et al. (2018). BiOX (X= Cl, Br, I) photocatalytic nanomaterials: Applications for fuels and environmental management. *Advances in Colloid and Interface Science, 254,* 76–93.
33. Zhai, W., & Zhou, K. (2019). Nanomaterials in superlubricity. *Advanced Functional Materials, 29*(28), 1806395.
34. Fang, L., et al. (2019). Turning bulk materials into 0D, 1D and 2D metallic nanomaterials by selective aqueous corrosion. *Chemical Communications, 55*(70), 10476–10479.
35. Wang, Z., et al. (2020). Application of zero-dimensional nanomaterials in biosensing. *Frontiers in Chemistry, 8,* 320.
36. Huang, D., et al. (2019). Megamerger in photocatalytic field: 2D g-C3N4 nanosheets serve as support of 0D nanomaterials for improving photocatalytic performance. *Applied Catalysis B: Environmental, 240,* 153–173.
37. Kufer, D., & Konstantatos, G. (2016). Photo-FETs: Phototransistors enabled by 2D and 0D nanomaterials. *ACS Photonics, 3*(12), 2197–2210.
38. Sahoo, S., Parveen, S., & Panda, J. (2007). The present and future of nanotechnology in human health care. Nanomedicine: Nanotechnology, *Biology and Medicine, 3*(1), 20–31.
39. Jones, R. (2004). The future of nanotechnology. *Physics World, 17*(8), 25.
40. Bhattacharya, S., et al. (2013). Role of nanotechnology in water treatment and purification: Potential applications and implications. *International Journal of Environmental Science and Technology, 3*(3), 59–64.

Chapter 4
Understanding the Physics
of Photocatalytic Phenomenon

Abstract The process of photocatalysis has received a lot of interest in recent years. The process of photocatalysis takes place in the presence of sunlight. The different catalysts are used during the process of photocatalysis like TiO_2, ZnO, metal oxides, nitrides, among others. The basic processes of photocatalysis have been addressed in detail in this chapter, with heterogeneous and homogeneous photocatalysis being the two primary forms of photocatalysis. Both types have their own advantages and disadvantages that can influence the catalytic performance of any semiconductor. A variety of factors can have a significant influence on a material's photocatalytic activity. The most important factors that affect photocatalysis are pH, type of semiconductor, doping, and recombination rate of electrons and holes.

4.1 Introduction

The term photocatalysis is first time used in the field of research in 1911. In the last few years, this process takes a lot of attention. The concept of photocatalysis comes from the process of photosynthesis [1]. The process of photocatalysis takes place in the presence of light, and it could be natural light or artificial light. In this process, the light interacts with the surface of the material, which is a semiconductor material and known as catalyst. As a result, two processes take place one is oxidation and the second is reduction. Because the catalyst is so critical in photocatalysis, semiconductors are employed in this process [2].

To begin with, TiO_2 has shown to be a highly effective catalyst. It has a band gap of 3.2 eV and absorbs more light during photocatalysis [3]. Other catalysts used in photocatalysis include ZnO, metal oxides, Fe_2O_4, CdS, ZnS, and others. In the photocatalysis process, nanomaterials such as silver, gold, copper, and nanoparticles of metal oxides, among others, are utilized as a more excellent and efficient photocatalyst. The photocatalyst, which is considered as a more efficient, has a band gap less or equal to 3.2 eV, and it also has high chemical and photostability [4]. TiO_2 is mostly used as a photocatalyst because it is non-toxic, environmentally friendly,

M. B. Tahir et al., *New Insights in Photocatalysis for Environmental Applications*,
SpringerBriefs in Applied Sciences and Technology,
https://doi.org/10.1007/978-981-19-2116-2_4

and low cost, along with the aforementioned properties. As a photocatalyst, one-dimensional nanotubes and nanofibers of TiO_2 are also employed. Photocatalysis is considered a more efficient process as compared to others processes because of its unique properties [5]. The photocatalysis process is used in environmental application, energy storage, conversion applications, etc. In environmental applications, it can be used in water purification, degradation of dyes, heavy metal ions reduction, and removal of other pollutants that are present in water. Because these pollutants can pollute the water and can directly affect human health. Water is the most essential element for the survival of life, so that's why purification of water is very important [6]. The process of photocatalysis is also used in the purification of air. The presence of hazardous gases in the air can degrade the freshness of the air, which is why air filtration is critical [7]. This method is also utilized in solar cells, lithium-ion batteries, and other batteries for energy storage and conservation. It has the potential to both create hydrogen and transform biomass into fuel. Photocatalysis is widely employed in a variety of sectors, and it produces good results when compared to other processes [8].

As recently, 2D materials have revolutionized material science and have won the hearts of the scientific community owing to their tremendous potential to serve as a solution to many problems, just after the invention of graphene in 2004 [9]. Their tensile strength, improved surface area, revolutionary magnetic, conductive, and electronic properties have made these materials a solution to a series of scientific problems. Several layered 2D materials have also been exfoliated from their bulk materials and have shown some surprising characteristics. Among all 2D materials such as graphene, transition metals dichalcogenides (MoS_2, $MoSe_2$, WSe_2), MXenes, and phosphorene monolayer proved to be rising stars in the field of photocatalysis [10].

4.2 The Basic Mechanism of Photocatalysis

The basic mechanism of photocatalysis is that a semiconductor is treated under ultraviolet (UV) or visible light and as a result a series of reactions occur [11]. Resultantly, light semiconductor generally absorbs the photons according to its band gap, and thus, the phenomena of excitation of electrons from the valence band (VB) to conduction band (CB) occur. As obvious from the fact that an electron, while leaving its original position during excitation, leaves a hole; thus, this electron–hole pair is often called photogenerated or photoinduced electron–hole (e/h) pair [12]. As illustrated in Fig. 4.1, the basic mechanism lies in the sequence of photoinduced charge pair formation leading to separation of this pair, which tries to diffuse towards the photocatalyst's surface, resulting in a redox reaction on active photocatalytic sites. That's how this process takes place [13].

The general equations for the photocatalytic mechanism are given below [14]:

$$Photocatalyst + hv \rightarrow photocatalyst \left(e^-(CB) + h^+(VB) \right) \quad (4.1)$$

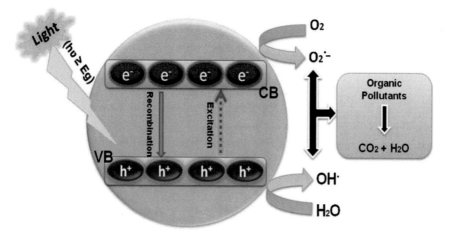

Fig. 4.1 Basic mechanism of photocatalysis [15]

$$H_2O + h^+(VB) \rightarrow OH^. + H^+ \tag{4.2}$$

$$O_2 + e^-(CB) \rightarrow O_2^- \tag{4.3}$$

$$O_2^{-.} + H^+ \rightleftharpoons HOO^. \tag{4.4}$$

$$2HOO^. \rightarrow H_2O_2 + O_2 \tag{4.5}$$

$$H_2O_2 \rightarrow 2OH \tag{4.6}$$

$$OH^. + organic\,pollutants \rightarrow CO_2 + H_2O \tag{4.7}$$

4.3 Heterogeneous Photocatalysis

In the process of heterogeneous photocatalysis, the reactants and catalyst are in different phases [16]. A large number of reactions take place for the dehydrogenation, detoxification of water, purification of air, deposition of metals, among others [17]. This catalysis can occur in several media like organic liquid phase, aqueous solution, or in the gas phase.

The excellent heterogeneous process can be taken place in different steps [18]:

- In the first step, the reactants are transferred to the surface.
- In the second step, at least one of the reactants is absorbed.
- In the third step, the reaction is in the adsorbed phase.
- In the fourth step, the desorption of the product takes place.
- In the last step, the products are removed from the interfacial area.

   The heterogeneous photocatalysis is efficiently used for the treatment of water and air [19]. By using this process, the pollutants are easily removed from the water and can remove toxic gases from the air [20]. The heterogeneous photocatalysis process is environmentally friendly, non-toxic, cost-effective, and can remove various types of pollutant including organic and inorganic.

4.4 Homogeneous Photocatalysis

In the process of homogeneous photocatalysis, the reactants and the catalyst exist in the same phase. The homogeneous photocatalysis process takes place in the liquid phase [21]. In industrial applications, homogeneous catalysts are used to accelerate the pace of a process without changing or increasing the temperature [22]. The ability to use sunlight up to 450 nm is the most crucial feature of this technique, since it eliminates the need for an extensive UV lamp. The disadvantage of this process includes low pH values [23].

4.5 Impact of Different Parameters on Photocatalytic Activity

Many different factors have a very important impact on the photocatalytic activity.

4.5.1 pH of Semiconductor

The pH of semiconductor affects photocatalytic activity. For example, if we take an example of photocatalytic degradation of dyes [24], in the photodegradation of dyes, the pH of the semiconductor is varied between 1 and 11. In every case, the value of pH is different, sometime efficient degradation of dye is obtained by varying the value of pH just at 3, and sometime by varying the value of pH at 10 or 11, the efficient degradation of dye is obtained in a short period of time [25].

   The process of photocatalytic oxidation also depends on the pH value of solution. If we use TiO_2 in the process of pollutant degradation, then TiO_2 gives efficient degradation of pollutants on the low value of pH [26]. The reason behind this is the surface of titanium dioxide is more positive in an acidic medium so, TiO_2 easily absorbs the pollutants. Increasing the value of pH can be shifted in the alkali region, so in this case, the surface of TiO_2 is negative, and then it can repel the pollutants or cannot absorb pollutants easily. So, it can affect the photocatalytic activity and also degradation efficiency [27]. The result is obtained that by increasing pH value, the efficiency of the degradation process is decreased. The production of hydroxyl radicals and hydroxide ions increases in an acidic media, improving photocatalytic activity. If nitrogen is removed during the photocatalytic oxidation process, the pH ranges from 1 to 11 [28].

4.5.2 Size of Semiconductor

The size of the semiconductor has a very important impact on photocatalytic activity. There are different sizes of semiconductors that are used in the photocatalytic activity. If the size of the semiconductor catalyst is in the micrometer range, then photocatalytic activity is not efficient or the rate of photocatalytic activity is low [29]. If the size of the semiconductor catalyst is in nanoscale, the photocatalytic activity is more efficient, and the rate of photocatalytic activity is fast. Because in microsize, the surface area of semiconductor is small, the atom on the surface is less, which slows down the kinetics of photocatalysis. The surface area of the semiconductor is large in nanoscale, resulting in a large number of atoms on the surface, which increases the rate of photocatalysis and makes the process more efficient [30].

4.5.3 Synthesis Techniques

The synthesis techniques also show an impact on photocatalytic activity [31]. There are many techniques used for the synthesis of semiconductors like hydrothermal method, sol–gel method, chemical method, coprecipitation method, microwave method, radiation method, direct oxidation method, solvothermal method, and electrodeposition technique. Because the synthesis process differs in each approach, distinct parameters are modified in each of these methods, resulting in a variable influence on photocatalytic activity. For example, in the hydrothermal method, we can easily control the temperature during the synthesis of semiconductor catalyst, so by varying the temperature, the morphology of the catalyst can be changed. So, with different morphology, the catalyst behaves differently during the photocatalytic activity, and this may affect the rate or the efficiency of the process. This is how the photocatalytic activity is affected by the synthesis procedure. **Recombination of charge carriers**.

The recombination of the charges also affects the photocatalytic activity. In the photocatalytic process, the charges must not be recombined fast, because the recombination of the charges affects the process of photocatalysis. In the photocatalysis process, when light falls on the surface of a semiconductor catalyst, the electron from the valance band goes to the conduction band and holes are generated in the valance band. Then the process of reduction and oxidation happens and hydroxyl radical and hydroxide ion is generated, and the process goes forward. The photocatalytic process cannot progress, and photocatalytic activity cannot be conducted if the electron that travels to the valance band returns rapidly. That's how it can affect photocatalytic activity. Different material is doped in the semiconductor catalyst, which helps them to control the fast recombination rate and enhancing the photocatalytic process.

4.5.4 Band gap

The band gap also affects the photocatalytic activity. The semiconductor catalyst used in the process of photocatalysis should possess a band gap 3.3 eV or less because it is important that the catalyst absorbed light efficiently during the process. If the absorption rate of the photocatalyst is high, the process of photocatalysis is very efficient. Moreover, if the photocatalyst absorbs more light, it can efficiently degrade the pollutant. If the absorption region of any photocatalyst is low, then different materials are doped in the photocatalyst to enhance its absorption region.

4.6 Conclusion

Photocatalysis is an extremely efficient approach for a variety of applications. It has been widely used in the last few years. Here, different type of catalyst is used, which can help to perform the process of photocatalysis efficiently. This process is widely used in environmental and energy applications, food storage, and agriculture, among others. It may be utilized in the environment for water purification, air purification, and energy storage devices, among other things. Furthermore, the process of photocatalysis has many advantages like non-toxic, fast, and environmentally friendly nature. The process can be divided into two categories—one is heterogeneous photocatalysis, and the second is homogeneous photocatalysis. Several factors affect the process of photocatalysis like band gap, recombination rate, size, surface area, synthesis techniques, pH, etc. Because of its exceptional features, photocatalysis has a very bright future.

References

1. Ameta, R., et al. (2018). Photocatalysis. *Advanced Oxidation Processes for Waste Water Treatment* (pp. 135–175). Elsevier.
2. Maeda, K., et al. (2006). Photocatalyst releasing hydrogen from water. *Nature, 440*(7082), 295–295.
3. Lee, S.-Y., & Park, S.-J. (2013). TiO2 photocatalyst for water treatment applications. *Journal of Industrial and Engineering Chemistry, 19*(6), 1761–1769.

4. Kudo, A. (2006). Development of photocatalyst materials for water splitting. *International Journal of Hydrogen Energy, 31*(2), 197–202.
5. Nakata, K., & Fujishima, A. (2012). TiO$_2$ photocatalysis: Design and applications. *Journal of Photochemistry and Photobiology C: Photochemistry Reviews, 13*(3), 169–189.
6. Dionysiou, D. D., et al. (2016). Photocatalysis: applications. *Royal Society of Chemistry*.
7. Ibhadon, A. O., & Fitzpatrick, P. (2013). Heterogeneous photocatalysis: Recent advances and applications. *Catalysts, 3*(1), 189–218.
8. Wenderich, K., & Mul, G. (2016). Methods, mechanism, and applications of photodeposition in photocatalysis: A review. *Chemical Reviews, 116*(23), 14587–14619.
9. Su, T., et al. (2018). Role of interfaces in two-dimensional photocatalyst for water splitting. *Acs Catalysis, 8*(3), 2253–2276.
10. Rahman, M. Z., et al. (2016). 2D phosphorene as a water-splitting photocatalyst: Fundamentals to applications. *Energy & Environmental Science, 9*(3), 709–728.
11. Saravanan, R., Gracia, F., & Stephen, A. (2017). Basic principles, mechanism, and challenges of photocatalysis. *Nanocomposites for Visible Light-Induced Photocatalysis* (pp. 19–40). Springer.
12. Zhang, J., et al. (2018). Mechanism of Photocatalysis. *Photocatalysis* (pp. 1–15). Springer.
13. Banerjee, S., et al. (2014). New insights into the mechanism of visible light photocatalysis. *The Journal of Physical Chemistry Letters, 5*(15), 2543–2554.
14. Zhang, J., & Nosaka, Y. (2015). Generation of OH radicals and oxidation mechanism in photocatalysis of WO3 and BiVO4 powders. *Journal of Photochemistry and Photobiology A: Chemistry, 303*, 53–58.
15. Irfan, S., et al. (2019). Critical review: Bismuth ferrite as an emerging visible light active nanostructured photocatalyst. *Journal of Materials Research and Technology, 8*(6), 6375–6389.
16. Pelizzetti, E., & Serpone, N. (2012). Homogeneous and Heterogeneous Photocatalysis (Vol. 174). Springer Science & Business Media.
17. Colmenares, J. C., & Xu, Y.-J. (2016). Heterogeneous photocatalysis. *Green Chemistry and Sustainable Technology*.
18. Bard, A. J. (1979). Photoelectrochemistry and heterogeneous photo-catalysis at semiconductors. *Journal of Photochemistry, 10*(1), 59–75.
19. Herrmann, J.-M., Guillard, C., & Pichat, P. (1993). Heterogeneous photocatalysis: An emerging technology for water treatment. *Catalysis Today, 17*(1–2), 7–20.
20. Palmisano, G., et al. (2010). Advances in selective conversions by heterogeneous photocatalysis. *Chemical Communications, 46*(38), 7074–7089.
21. McCormick, T. M., et al. (2010). Reductive side of water splitting in artificial photosynthesis: New homogeneous photosystems of great activity and mechanistic insight. *Journal of the American Chemical Society, 132*(44), 15480–15483.
22. Cieśla, P., et al. (2004). Homogeneous photocatalysis by transition metal complexes in the environment. *Journal of Molecular Catalysis A: Chemical, 224*(1–2), 17–33.
23. Oller, I., et al. (2007). Solar heterogeneous and homogeneous photocatalysis as a pre-treatment option for biotreatment. *Research on Chemical Intermediates, 33*(3–5), 407–420.
24. Fernández, J., et al. (2004). Orange II photocatalysis on immobilised TiO$_2$: Effect of the pH and H$_2$O$_2$. *Applied Catalysis B: Environmental, 48*(3), 205–211.
25. Rincon, A. -G., & Pulgarin, C. (2004). Effect of pH, inorganic ions, organic matter and H2O2 on E. coli K12 photocatalytic inactivation by TiO$_2$: Implications in solar water disinfection. *Applied Catalysis B: Environmental, 51*(4), 283–302.
26. Chu, W., Choy, W., & So, T. (2007). The effect of solution pH and peroxide in the TiO$_2$-induced photocatalysis of chlorinated aniline. *Journal of Hazardous Materials, 141*(1), 86–91.
27. Bhatkhande, D. S., Pangarkar, V. G., & Beenackers, A. A. C. M. (2002). Photocatalytic degradation for environmental applications–A review. *Journal of Chemical Technology & Biotechnology: International Research in Process, Environmental & Clean Technology, 77*(1), 102–116.
28. Merka, O., et al. (2011). pH-control of the photocatalytic degradation mechanism of rhodamine B over Pb3Nb4O13. *The Journal of Physical Chemistry C, 115*(16), 8014–8023.

29. Stroyuk, A., et al. (2005). Quantum size effects in semiconductor photocatalysis. *Theoretical and Experimental Chemistry, 41*(4), 207–228.
30. Li, Y.-F., & Liu, Z.-P. (2011). Particle size, shape and activity for photocatalysis on titania anatase nanoparticles in aqueous surroundings. *Journal of the American Chemical Society, 133*(39), 15743–15752.
31. Dodoo-Arhin, D., et al. (2018). The effect of titanium dioxide synthesis technique and its photocatalytic degradation of organic dye pollutants. *Heliyon, 4*(7), e00681.

Chapter 5
Role of Metal Oxide/Sulphide/Carbon-Based Nanomaterials in Photocatalysis

Abstract The current state of pollution in the ecosystem is the most serious warning for humans. Many techniques have been used for the removal of toxic pollutants from the environment, but photocatalysis came out as a most promising technique among all of the methods. Photocatalysis can be used for the removal of dyes, splitting of water through reduction and oxidation. Different semiconductors are used as photocatalysts where metal oxides, metal sulphides, and carbon-based materials are emerged as more efficient photocatalysts. Oxides are used for their unique properties such as high stability, non-toxic by-products, high abundance, and demonstrate excellent behaviour towards the UV light and visible light. Sulphides are utilized as photocatalysts because of their acceptable band gaps, which are crucial in the field of photocatalysis. Nanosized carbon base materials lead us to many benefits such as wide surface area, thermal stability, numerous functional groups, small resistance, and large pore size. Carbon nanomaterials are used in pollutants adsorption from water, composting, soil, and in photocatalysis. This chapter covers the limitation of various semiconductors towards the photocatalytic behaviour.

5.1 Introduction

Environmental pollution has raised serious health concern. The cause of pollution is industrialization and modern transportation, which emit toxic organic compounds such as hazardous gases, dyes, greenhouse gases, pesticides, and volatile organic compounds. For the proper risk-free growth of environment, we need a well-organized method to remove these toxic organic pollutants. Photocatalysis is the most dominating technique among all the decomposition methods due to its inherent properties. During the process of photosynthesis, all the photocatalyst substances are treated under the UV light and electron hole pair is generated, which is helpful in the redox reaction. These active radicals are important in the photocatalytic activity or degradation of toxic pollutants, in which electron hole pairs operate as reactive agents for oxidation and reduction, allowing poisonous chemicals to be removed [1]. Over the recent decade, there are numerous semiconductor materials that are used as

© The Author(s), under exclusive license to Springer Nature Singapore Pte Ltd. 2022   39
M. B. Tahir et al., *New Insights in Photocatalysis for Environmental Applications*,
SpringerBriefs in Applied Sciences and Technology,
https://doi.org/10.1007/978-981-19-2116-2_5

photocatalysts. The most common photocatalysts are metal oxides, metal sulphides, and carbon-based materials because of their promising characteristics.

There are a number of metal oxides and alkali metals used in the removal of bacteria, fungi, cancer cells, and purification of water and air. Oxides such as ZnO, WO_3, CuO, and TiO_2 are mostly used in photocatalysis. However, TiO_2 demonstrates promising photocatalytic characteristics. These metal oxides are used under the UV light which is a small part (only 4%) of the electromagnetic spectrum. The large band gap and recombination of electron/hole pair are the main problems of the metal oxides [2]. While metal sulphides have a small band gap so they are more suitable for the process of photocatalysis. Due to their short band gap, they can proceed with light directly and easily. The positions of bands made them more promising for the absorption of light. Because of its elevated location, the conduction band can readily absorb UV radiation and be used for the reduction of water. Because of their high absorption capacity, sulphides are poor candidates for direct water splitting. So, there is still a need for another promising photocatalytic agent for the splitting of water. To answer the aforementioned questions, sulphides performed well when there is a liquid solution such as Na_2S and Na_2SO_3 [3].

Carbon-based materials are also used for the contamination of dyes and pollutants from the liquids. Nanosized carbon materials are more suitable for the photocatalytic process. Because of its elevated location, the conduction band can readily absorb UV radiation and used for the reduction of water. Because of their high absorption capacity, sulphides are poor candidates for direct water splitting [4].

Metal oxides and sulphides have different chemical formulas, making metal sulphides the best choice as the photocatalyst and catalytic material for different energy applications such as batteries, water splitting, and hydrogen production catalytically. Since sulphides nanostructures have unique characteristics such as better redox potential, high sensitivity, precise capacity, and more lifetime than oxides, metal sulphides also have low melting point than oxides. Thus, in energy conversion applications, oxygen is replaced by sulphur provides more stable, sensitive, and selective photosensing material. Metal sulphides have a lower melting point than oxides. While metal oxides offer a variety of properties, such as high sensitivity, low cost, and ease of integration, they are most commonly used in electronics [5].

5.2   Metal Oxide-Based Nano Photocatalysts

Metal oxides are the cost-effective and ecofriendly photocatalysts that help to decompose waste and toxic pollutants because owing to their physical, electronic, and chemical properties. Metal sulphides have a lower melting point than oxides. While metal oxides offer a variety of properties, such as high sensitivity, low cost, and ease of integration, they are most commonly used in electronics. All the materials that have to be used as photocatalysts should absorb light, and they should have stable morphology. Moreover, the materials should have a higher surface area and should be able to be recycled. Metal oxides are frequently utilized in industries and have a variety of applications due to their good physical, chemical, and electrical characteristics. TiO2 is commonly employed as a photocatalyst in both its pure and doped forms on a wide scale. There is a broad role of TiO_2 in the applications of purification and degradation of water, air, and soil. The glass based on TiO_2 has the ability to clean itself. There is another catalyst named ZnO, which is a very important catalyst for photocatalysis. When other metals are doped with ZnO, the applications are widened in the field of photocatalysis [6]. Iron-based photocatalysts are also employed in biodegradation, and Fe2O3 can be used in photo-electrolysis cells. MgO photocatalysts have a significant band gap and can be utilized to degrade dyes in liquids. When silver and gold are doped with the tungsten oxide WO_3, then a more efficient photocatalytic process occurs. The nano-oxide of Cu has a narrow band gap and utilizes in the purification of water under visible light. All oxide-based photocatalysts can be handled under ultraviolet and visible light depending on the band gap of the photocatalyst semiconductor. All the oxide-based photocatalysts depend on the size, shape, composition, and morphology [7].

5.2.1 Oxides of Group-IV-Based Photocatalyst

Titanium oxide (TiO_2) is the most commonly used semiconductor photocatalyst for the splitting of water. This photocatalyst has three different phases of crystal which are stable at environmental condition. The first phase of crystal is the anatase phase which is a meta-stable crystal, and this is excellent for water splitting and photocatalysis having a large band gap of 3.2 eV. The second form is rutile which is stable in thermodynamic terms whose band gap is narrower as compared to the anatase phase and the last form is brookites phase which is highly pure and reusable. The brookite phase is not famous in photocatalysis because of its physicochemical characteristics.

Titanium dioxide can be fabricated in nanobelts, rods, wires, and nanofibers. It is a highly stable, cost-effective, and extra efficient catalyst for water splitting and generation of H_2. It can be used for PEC splitting of water applications. Nowadays, the researchers focus on practically achieving high quantum efficiency, highly efficient light harvesting, and stability. Different synthesis methods are used to increase the photocatalytic process efficiency by using TiO_2 as a photocatalyst [8].

Recent research has concentrated on improving the effectiveness of photocatalytic processes when exposed to visible light. By mixing the chemicals by ultrasonication method, hybrid photocatalysts can be fabricated and no covalent bond exists. So, there would be no interaction between the elements participating in reactions.

Pure TiO_2 is comparatively less efficient for the photocatalytic process as compared to the composite of TiO_2. So, different methods such as sol–gel, hydrothermal, solvothermal, as well as many other methods are used for the fabrication of composite or doped TiO_2 [9].

5.2.2 Oxides of Group-V-Based Photocatalyst

Bismuth belongs to fifth group of periodic table. The semiconductor made up of oxides is very reactive under visible light and displays very effective photocatalytic property. The electronic structure is $2p$ for oxygen and $6s$ for bismuth displayed by valence band. The valence band of bismuth shows $6s$ structure, which means a decreased band gap and becomes reason for mobility of carriers enhanced by $6s$ orbitals. All material based on bismuth have a band gap, which is less than 3 eV. Semiconductor based on bismuth has applications in the field of dye degradation and removing the waste of water. Bismuth oxides are widely used in the sensors as well as in fuel cells which are constructed by the ceramic glass. The band of bismuth-based oxides lies between 2.4 and 2.8 eV which is very responsive to visible light. The chemical instability of bismuth oxide is a major drawback. The photocatalytic activity can be improved by doping and composites. Under visible light, platinum-coated bismuth oxides exhibit exceptional photocatalytic activity and function as a highly efficient reactor [10].

Vanadium oxide (V_2O_5) is a modern oxide that may be used to oxidize and reduce hydrocarbons like NOx and NH_3, as well as for gas monitoring and Li-ion batteries. In recent decades, the photocatalytic performance of V_2O_5 was observed by different scientists and compounds based on V, O are widely used in dye degradation and water splitting. This shows that V_2O_5 is a promising photocatalyst. The photocatalytic activity can be enhanced by making a composite of Al_2O_3 with V_2O_5. The V, Al ratio, their temperature and heating duration, and the photocatalytic performance of the material are all factors to consider. The enhancement of photocatalytic activity depends on the band structure and redox potentials [11].

5.2.3 Oxides of Group-VI-Based Photocatalyst

Tungsten-based oxides are cost-effective and show excellent photocatalytic properties just because of their band gap, that is why they are used on a large scale in different applications such as solar energy, electrochromic devices, and chemical sensors. For the composition of the WO_3 structure, several chemical processes are utilized, but electrodecomposition has gained popularity due to its unique qualities, such as flexibility in increasing the surface area and cost efficiency. Tungsten oxides are used on a large scale because they are stable and have affordable costs [12].

It has a large band gap so it can utilize only 3–5% of ultravisible light. But small band gap semiconductor can perform much better than a large band gap so, there is need of a catalyst which can perform better under visible light such as WO_3 which have band gap of 2.4–2.8 eV. However, WO_3 has high coercivity as well as high adsorption rate. Furthermore, WO_3 possesses physical and chemical stability, but TiO_2 has a low conductivity, which is why it has limited applicability in electronics [13].

5.2.4 Oxides of Group-VII A-Based Photocatalyst

All the properties of nanomaterials depend on the chemical composition as well as morphology of the material. The different fabrication methods can change the property of material on basis of shape, size, and procedure. Various synthesis methods are used for the fabrication of manganese oxides and hydro-oxide nanostructures. Because of its cost-effectiveness and ease of operation, the hydrothermal process may be utilized to fabricate manganese oxides. It can proceed on low temperature which leads to the controllable size of particles.

Different chemical methods including hydrothermal method, electro decomposition, microwave flux, and ultrasonic wave process are used to fabricate the manganese oxides [14].

5.2.5 Oxides of Copper and Zinc

Zinc oxide is the element of II-IV groups and is a promising photocatalyst because of its cost-effectiveness, availability, and ecofriendly properties. This semiconductor is widely used for the degradation of dyes during the catalytic procedure. The photocatalytic activity of ZnO can be improved by different methods. Firstly, the material should be targeted with visible light instead of ultraviolet light because visible light can generate more electron hole (e/h) pairs. Secondly, the controlled process of photocatalysis can lead to the continuous use of photocatalyst in industries. Last and third one is the controlled morphology of ZnO, which is a key factor in the enhancement of the photocatalytic process [15].

The nanostructure of copper oxides is widely used at the industrial level. There are two types of p-type semiconductors, which include CuO and Cu_2O. Where CuO is monoclinic, small band gap of 1.2 eV and Cu_2O is cubic in shape and have band gap of 2.1 eV. They have a lot of applications in solar cells, photocells, field emitters, and superconducting materials. Both types of copper oxides can be used as photocatalysts because of their sizes; if Cu < 3 nm, it can efficiently produce charge carriers which leads to the reduction of oxygen. Copper oxide nanoparticles are employed in a variety of applications, including water splitting and hydrogen synthesis via photocatalysis [16].

5.3 Metal Sulphides as Nanophotocatalyst

Metal sulphides are more efficient photocatalysts than metal oxides due to their much small band gaps and their different physiochemical properties. The fast recombination of electron/hole pair is the major drawback of metal sulphides, making their use limited in photocatalysis. Many reports have been published to solve this problem by using various techniques and forming the different heterostructure. It was also reported that the formation of heterostructure could overcome the problem of photo-corrosion and recombination of carriers. The fabrication of junctions promotes photo-excited electrons. By increasing the surface area or by doping the different metals into the sulphides can increase the efficiency of photocatalytic performance. So, sulphides can be used as catalysts widely because of their efficient behaviour towards the excitation of electrons. Many researchers are looking at non-toxic/environmentally friendly metal sulphides and catalysts that do not contain noble metals [17].

Usually, metal sulphides have edge because of defective morphologies, different number of layers, and high crystallinity, which is essential for better photocatalytic performance. There is a large contribution of metal sulphide in the field of photo-catalysis for a decade. Metal sulphides have attracted attention due to their potential properties, which include active sites, good catalytic and photocatalytic capabilities, a narrower energy band gap, and tiny masses. The crystalline structure of sulphides is easily available in the market and shows best behaviour towards the light, making them an excellent photocatalyst in terms of practical application. A number of metal sulphides and their nanostructures are used in energy and storage applications [18].

5.3.1 CuS as Nanophotocatalyst

The composite of CuS/ZnS is the most precise photocatalysts for the evaluation of hydrogen. Wang et al. reported the generation of H_2 under the induced visible light by using the composite of CuS/ZnS. By increasing the copper ions Cu^{2+} the generation rate of H_2 can be increased. The generation rate of H_2 can be increased to a limit. When the Co catalyst copper ions are reached at its maximum limit, hydrogen will be evaluated. So, it is clear that the generation rate can be increased to a specific limit of 7 mol % [19].

5.3.2 MoS₂ as Nanophotocatalyst

In the photocatalysis process, several metal sulphides are utilized as photocatalysts, but MoS_2 stands out due to its excellent stability, superb electrical arrangement, and photo-absorption properties. As MoS_2 has band gap of 1.2 eV which is not suitable for any photocatalysis reaction, the oxidation and reduction process of MoS_2 can help start the photocatalytic process. While 2D nanosheets have a quantum limitation effect as they have suitable band gap of 1.96 eV [20].

5.3.3 ZnS-Based Photocatalysts

ZnS has a large band gap of 3.6 eV that can operate under UV light; due to the high band position of ZnS, it can lead to a high quantum yield. Many studies are reported regarding to the use of visible light by keeping the conduction band high. A lot of doping materials are used for the enhancement of photocatalytic activity by decreasing the band gap which can absorb more light. It is reported by researchers that the photocatalytic activity of ZnS is increased after doping with different dopants [21]. ZnS retains a high conduction band after doping, resulting in outstanding photocatalytic activity without the use of a cocatalyst. When the dopants are non-metals, results are outstanding, showing excellent behaviour towards the visible light. When ZnS is mixed with CuS, it shows excellent response in visible light and shows excellent photocatalytic activity. When the transformation of charge occurs from ZnS to CuS, then CuS could be reduced to Cu_2S [22].

5.3.4 Carbon Nanotubes as Cocatalysts

Single-wall carbon nanotubes (SWCNTs) drag attention due to unique properties including ballistic transport, mobilization, and high density of current; hence, they play an important role in electronics. The fabrication of SWCNTs should be less than 400 °C for escaping from the dielectric films produced during the degradation of SWCNTs' growth. Many researches have been reported the fabrication of SWCNTs on low temperature. Recently, a report has been made on the fabrication of SWCNTs under 300 °C. Ni, Co, and Fe are the metal catalysts that are really very important in different practical applications [23].

5.3.5 Carbon-Based 2D Materials

Carbon-based materials are widely used in different applications including batteries, fuel cells, and supercapacitors as well as in photocatalysis because of their affordable costs and non-toxic behaviour. Graphene is an important carbon-based materials due to its physiochemical properties. Graphene is widely used for the CO_2 reduction. The large surface area, thermally conducting behaviour and chemically stable graphene, leads to the photocatalytic activity of CO_2 reduction. It can increase the photocatalytic performance during the removal of CO_2 [24].

The $g-C_3N_4$ is another carbon-based 2D material which can be used for improving the energy and environmental applications. The $g-C_3N_4$ has an intermediate band gap of 2.7 eV, showing that $g-C_3N_4$ is an excellent semiconductor while graphene is a good conductor. The reduction of CO_2 is done by using the $g-C_3N_4$ semiconductor.

The g-C_3N_4 is a good semiconductor, it shows many properties such as better absorption of light as compared to other common semiconductors. It also has a large surface area, it is chemically highly stable and the cost-effectiveness is another good property of the g-C_3N_4 semiconductor. All these properties made them stable and more attractive for the reduction of CO_2 by using photocatalytic activity [25].

5.4   Conclusion

This chapter covers the precise study of metal oxides, sulphides, and carbon-based materials as photocatalysts in the photocatalysis process as well as their applications and advantages in the environmental applications. The fast recombination of carriers in sulphides can be solved by using the different materials as doping materials. All these semiconductor materials show good behaviour under visible light as well as UV-light during the process. Metal oxides show different photocatalytic behaviour due to their different morphology and band gaps. By doping the other materials, the photocatalytic activity of semiconductor photocatalyst will be enhanced.

References

1. Ameta, R., Solanki, M. S., Benjamin, S., & Ameta, S. C. (2018). Photocatalysis. In *Advanced Oxidation Processes for Waste Water Treatment* (pp. 135–175). Academic Press.
2. Karimi-Maleh, H., Kumar, B. G., Rajendran, S., Qin, J., Vadivel, S., Durgalakshmi, D., Gracia, F., Soto-Moscoso, M., Orooji, Y., & Karimi, F. (2020). Tuning of metal oxides photocatalytic performance using Ag nanoparticles integration. *Journal of Molecular Liquids, 314*, 113588.
3. Chandrasekaran, S., Yao, L., Deng, L., Bowen, C., Zhang, Y., Chen, S., Lin, Z., Peng, F., & Zhang, P. (2019). Recent advances in metal sulfides: From controlled fabrication to electrocatalytic, photocatalytic and photoelectrochemical water splitting and beyond. *Chemical Society Reviews, 48*(15), 4178–4280.
4. Brião, G. V., de Andrade, J.R., da Silva, M. G. C., & Vieira, M. G. A. (2020). Removal of toxic metals from water using chitosan-based magnetic adsorbents. A review. *Environmental Chemical Letters*. https://doi.org/10.1007/s10311-020-01003-y
5. Yadav, P., Dwivedi, P. K., Tonda, S., Boukherroub, R., & Shelke, M. V. (2020). Metal and non-metal doped metal oxides and sulfides. In *Green Photocatalysts* (pp. 89–132). Springer.
6. Kato, Y., Yamamoto, M., Akatsuka, M., Ito, R., Ozawa, A., Kawaguchi, Y., Tanabe, T., & Yoshida, T. (2019). Study on carbon dioxide reduction with water over metal oxide photocatalysts. *Surface and Interface Analysis, 51*(1), 40–45.
7. Kusior, A., Michalec, K., Jelen, P., & Radecka, M. (2019). Shaped Fe_2O_3 nanoparticles–synthesis and enhanced photocatalytic degradation towards RhB. *Applied Surface Science, 476*, 342–352.
8. Moss, B., Lim, K. K., Beltram, A., Moniz, S., Tang, J., Fornasiero, P., Barnes, P., Durrant, J., & Kafizas, A. (2017). Comparing photoelectrochemical water oxidation, recombinationkinetics and charge trapping in the three polymorphs of TiO_2. *Science and Reports, 7*, 2938.
9. Gupta, B., & Melvin, A. A. (2017). TiO_2/RGO composites: its achievement and factors involved in hydrogen production. *Renewable and Sustainable Energy Reviews, 76*, 1384e92.

10. Pei, C. C., Lo, K. K. S., & Leung, W. W. F. (2017). Titanium-zinc-bismuth oxides-graphene composite nanofibers as high-performance photocatalyst for gas purification. *Separation and Purification Technology, 184*, 205–212.
11. Wang, S., Li, Z., Guan, Y., Lu, L., Shi, Z., Weng, P., Yan, S., & Zou, Z. (2019). Visible light driven $TaON/V_2O_5$ heterojunction photocatalyst for deep elimination of volatile-aromatic compounds. *Applied Catalysis B: Environmental, 245*, 220–226.
12. Jeevitha, G., Abhinayaa, R., Mangalaraj, D., & Ponpandian, N. (2018). Tungsten oxide-graphene oxide (WO3-GO) nanocomposite as an efficient photocatalyst, antibacterial and anticancer agent. *Journal of Physics and Chemistry of Solids, 116*, 137–147.
13. Zhang, X., Hao, W., Tsang, C. S., Liu, M., Hwang, G. S., & Lee, L. Y. S. (2019). Psesudocubic phase tungsten oxide as a photocatalyst for hydrogen evolution reaction. *ACS Applied Energy Materials, 2*(12), 8792–8800.
14. Atacan, K., Topaloğlu, B., & Özacar, M. (2018). New CuFe2O4/amorphous manganese oxide nanocomposites used as photocatalysts in photoelectrochemical water splitting. *Applied Catalysis A: General, 564*, 33–42.
15. Durmus, Z., Kurt, B. Z., & Durmus, A. (2019). Synthesis and characterization of graphene oxide/zinc oxide (GO/ZnO) nanocomposite and its utilization for photocatalytic degradation of basic Fuchsin dye. *ChemistrySelect, 4*(1), 271–278.
16. Sorbiun, M., Mehr, E. S., Ramazani, A., & Fardood, S. T. (2018). Green synthesis of zinc oxide and copper oxide nanoparticles using aqueous extract of oak fruit hull (jaft) and comparing their photocatalytic degradation of basic violet 3. *International Journal of Environmental Research, 12*(1), 29–37.
17. Lee, S. L., & Chang, C. J. (2019). Recent Progress on metal sulfide composite nanomaterials for photocatalytic hydrogen production. *Catalysts, 9*(5), 457.
18. Weidner, J. W. (2018). Solar energy: An enabler of hydrogen economy? *Electrochemical Society Interface, 27*(1), 45.
19. Zong, X., Yan, H., Wu, G., Ma, G., Wen, F., Wang, L., & Li, C. (2008). Enhancement of photocatalytic H_2 evolution on CdS by loading MoS_2 as cocatalyst under visible Light irradiation. *Journal of the American Chemical Society, 130*(23), 7176–7177.
20. Chang, C. J., Huang, K. L., Chen, J. K., Chu, K. W., & Hsu, M. H. (2015). Improved photocatalytic hydrogen production of ZnO/ZnS based photocatalysts by Ce doping. *Journal of the Taiwan Institute of Chemical Engineers, 55*, 82–89.
21. Lee, G. J., & Wu, J. J. (2017). Recent developments in ZnS photocatalysts from synthesis to photocatalytic applications—A review. *Powder Technology, 318*, 8–22.
22. Ye, Z., Kong, L., Chen, F., Chen, Z., Lin, Y., & Liu, C. (2018). A comparative study of photocatalytic activity of ZnS photocatalyst for degradation of various dyes. *Optik, 164*, 345–354.
23. Okada, T., Saida, T., Naritsuka, S., & Maruyama, T. (2019). Low-temperature synthesis of single-walled carbon nanotubes with Co catalysts via alcohol catalytic chemical vapor deposition under high vacuum. *Materials Today Communications, 19*, 51–55.
24. Innocenzi, P., & Stagi, L. (2020). Carbon-based antiviral nanomaterials: Graphene, C-dots, and fullerenes. A perspective. *Chemical Science, 11*(26), 6606–6622.
25. Weng, G. M., Xie, Y., Wang, H., Karpovich, C., Lipton, J., Zhu, J., Kong, J., Pfefferle, L. D., & Taylor, A. D. (2019). A promising carbon/g-C_3N_4 composite negative electrode for a long-life sodium-ion battery. *Angewandte Chemie, 131*(39), 13865–13871.

Chapter 6
Plasmonic Photocatalysts and Their Applications

Abstract Photocatalysis has been accepted as a promising technique worldwide to resolve environmental issues, especially pollution and energy crisis. Photocatalysis has provided pathways in the field of electronics by the use of nanoparticles of semiconductors. However, the limitation in using semiconductor catalysts can minimize overall photocatalytic performance. Research has shown that a narrow band gap is suitable for visible light irradiation, but the material in that region may lack instability. On the other hand, a wide band gap semiconductor may be more stable, but it can only absorb ultraviolet (UV) light. To resolve this issue, many solutions are proposed including the elemental doping, composite cocatalysts, and heterostructure. Metal loading on the surface of a semiconductor can also provide the solution to deal with the recombination rate of charge pairs. Using plasmonic materials in reaction with other semiconductors can efficiently achieve a variety of environmental applications connected to photocatalysis.

6.1 Introduction

Plasmonic materials in interaction with other semiconductors can be used to efficiently achieve a number of photocatalysis-related environmental applications. The amount of harmful contaminants in the environment is growing as industrialization spreads. There is a list of contaminants, approximately 200 were reported in China, which are very dangerous for the aquatic system because of their toxicity and are also hazardous for human health. So, the exclusion of the toxic pollutants and wastes were the main problem of the present time. The control of the wastes was the biggest challenge; for this purpose, a number of techniques can use for the removal of dyes, including osmosis, membrane filtration, ion exchange, chlorination and electrochemical methods, are also reported in the literature. All these techniques are highly inefficient as well as expensive. The inefficiency and high cost of these techniques are major drawbacks. Photocatalysis was the technique that came to light due to its favourable characteristics. It inspired many scientists after resolving some energy issues and splitting of water and utilizing for the removal of dyes [1].

© The Author(s), under exclusive license to Springer Nature Singapore Pte Ltd. 2022   49
M. B. Tahir et al., *New Insights in Photocatalysis for Environmental Applications*,
SpringerBriefs in Applied Sciences and Technology,
https://doi.org/10.1007/978-981-19-2116-2_6

Photocatalysis is a modern technique that is utilizing for removing dyes, waste material, splitting of water, controlling environmental pollution, and generating different fuels. Photocatalysis has very effective characteristics such as better efficiency, cost-effectiveness, and easy to handle technique. It's a simple, sophisticated, and effective process for removing colours and harmful pollutants. This method, which employs oxygen as a chemical molecule, enables for the elimination of pollutants while simultaneously creating harmless by-products [2].

The simple photocatalysis mechanism involves a set of steps in which photons from light fall on the surface of the semiconductor. Then electrons in the valence band of that semiconductor get excited and go to the conduction band, leaving a hole in the valence band. The photogenerated hole and electron pairs are further involved in the process of oxidation and reduction at valence band and conduction band, respectively. As a result, the hydroxyl group and oxygen free radicals are produced, which helps to degrade pollutants and some by-products are also produced [3].

Different semiconductors are employed for this purpose, but poor adsorption region, least visible light absorption, quick rate of recombination, and least surface coverage are some of the drawbacks of semiconductor catalysts [4].

Photocatalysts can covert fuel on a very large scale and serve as a fundamental source of direct conversion of solar energy into chemical energy. In this way, carbon dioxide can also be reduced through water splitting. However, no such catalyst with the degradation performance is greater than 10% under single solar radiation. This standard is for any commercial level photocatalyst. However, experimentally it is impossible to achieve this on a large scale due to the recombination rate of electron hole pair. A narrow band gap can help in absorbing more visible light. The required voltage to split water is between 1.9 and 2.3 eV. However, the other issue with a narrow band gap is that it lacks stability during reaction under UV-visible light. In contrast, wide band gap semiconductors can provide better stability under UV. This can be reduced through heterojunction, composites, and different metals loading (plasmonics). Figure 6.1 explains three functional mechanisms of plasmonic-based photocatalysis. As it can be seen that the incorporation of plasmonic metal can enhance light absorption. After which redox (reduction and oxidation) reaction occurs and electrons go from valence band (VB) to conduction band (CB) of the semiconductor. E.F is the Fermi level at equilibrium and E vacuum is the vacuum energy level [5].

By the pairing of plasmonic nanostructure along with semiconductors, heterojunctions photocatalysts are prepared. These heterojunctions can increase photocatalytic activity by transforming energy from metal to semiconductor. The plasmonics can enhance the property of absorption of light. The coating of plasmonics on the surface of the semiconductor may boost the absorption rate by acting as an adjustable transmitter for the solar spectrum. Plasmonics has only recently been discovered and proved as a good photocatalysts. There are three different fields for the transformation of energy from a metal to semiconductor which are given below.

i.   Light scattering/trapping.

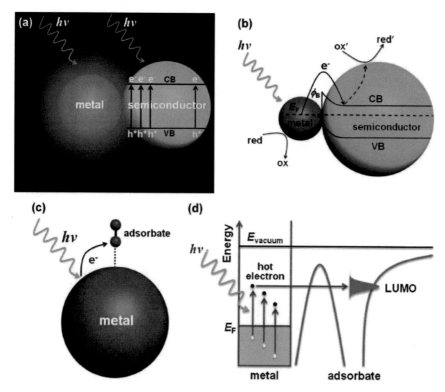

Fig. 6.1 Schematics of plasmonic based chemical reactions **a** enhanced light absorption by plasmonic, **b** plasmonic sensitization, **c** and **d** pure-metal plasmonic photocatalysis, permission taken from Ref. [6] published by the permission of Royal Society of Chemistry

ii.   Plasmon-induced resonance energy transfer (PIRET).
iii. Hot electron injection [7].

6.2 Functional Mechanism of Enhanced Plasmonic

There are two groups to describe plasmonic enhancement chemical reactions based on the type of catalysts being used. The first group is known as a hybrid group that can be described as an active semiconductor catalyst with the incorporation of metal plasmon nanoparticle ingredients. If greater light absorption is considered for any semiconductor, then just the surface of the semiconductor will be the major host for chemical reaction for higher localized surface plasmonic resonance (LSPR) effect. Moreover, for any wide band gap semiconductor plasmonic visible light irradiation can be made possible through plasmonic sensitization. Plasmonic sensitization is the transfer of electrons due to the excitation of LSPR that occurs at the interface of metal and semiconductor. This makes both holes in metal and electrons in semiconductor able to perform the photocatalytic activity.

The second group is named as direct catalytic effect, which occurs directly under resonant-excitation on metal (plasmonic) surface. This is also named as pure-metal plasmonic photocatalysis. The processes in this form of photocatalysis are mostly driven by local plasmonic heating. This reaction is further assisted by the injection of energetic electrons through plasmon excitation in vacant molecular orbitals within adsorbate [8]. Plasmon-enhanced mechanism is based on three functional steps which are discussed in detail below.

6.2.1 Plasmonic Enhancement to Absorb Light

Amplified optical signals with higher strength can arise in molecular species through the near field enhancement effect caused by localized plasmons. Such an effect can lead to the overlapping of spectral absorption of semiconductor with the nanocrystals of plasmonic metal resonance. There is a hybrid nanostructure that is fabricated by the overlap of semiconductor absorption with plasmonic metal resonance. The LSPR from a metal nanostructure is stimulated in this case by incident light, which causes simultaneous electron excitation and hole creation, resulting in a photogenerated pair. Only the part of semiconductor near metal surface will face the strong local field. The intensity of that field is typically far higher than the intensity of the far-field incident photon. The plasmonic excitation-induced amplification of the electric field tends to diminish exponentially away from the metal's surface. However, on the basis of size and shape of nanocrystals, the electric field enhancements extend from 10 to 50 nm [9]. Due to this range of spaces between nanocrystals, the rate of production of photogenerated electron–hole pairs can be increased in a semiconductor. However, plasmonic resonance energy can also be used to enhance the production rate of photoinduced pair.

The plasmonic enhanced absorption of light approach primarily uses metal nanocrystals as a light energy processor. Once then, due to near field interaction, the LSPR is transferred to the near optical species after the incoming photons' energy is highly concentrated. To apply photoelectrode in water splitting, the light absorption can be enhanced by the effect of plasmonic concentration of light, in thin film. This overcomes the carrier's limited diffusion lengths by altering the thickness of the absorption layer. Moreover, another factor that increases the absorption in semiconductors due to plasmonic is the greater cross-sectional area for scattering incident photons. This is because repeated scattering of incoming light in the presence of plasmonics has a significant impact on the optical path length, resulting in efficient absorption. For the enhancement in a semiconductor by using plasmonic, the following three factors should be overlapped with one another regardless of linear or nonlinear semiconductor absorption:

1. Light illumination source.

2. Semiconductor absorption.
3. LSPR of plasmonic metal nanocrystals [10].

However, some specific geometrical and spatial parameters of both metal and semiconductor should also be considered. This is because of the relation between the shape and space of metal nanoparticles and the local field induced by plasmonic excitation. Figure 6.1a has explained the phenomena of plasmonic enhancement light absorption.

6.2.2 Plasmonic Sensitization

This phenomenon is basically the separation of charges driven by plasmon at a wide band gap semiconductor–metal interface. Mostly, N-type metal oxides are used as photocatalysts Owing to \their contact with plasmonic metal bends down the conduction band at the thermal equilibrium of Fermi level. This process creates a potential barrier known as the Schottky barrier, which prevents electron transport between metal and semiconductor. This will generate the concentration of electrons that are not in equilibrium (thermal) with metal lattice under high resonance excitation. Only the electrons having higher energy than Schottky barrier will be able to cross that barrier. So, that fraction of electrons will induce in the conduction band of semiconductor. This charge transfer is same as the transfer process in dye-sensitized solar cells. However, the direction for the transfer of electrons in this case is opposite to the conventional semiconductor-based photocatalysts. Plasmonic sensitization is the transfer of electrons from a semiconductor to a metal and vice versa under different light excitations. On excitation of a semiconductor, electrons absorb energy greater than the band gap to get excited from semiconductor to metal. However, when the plasmon excitation occurs in metal, it excites the metal and then the hot electrons can cross the Schottky barrier to get entered in semiconductor conduction band. Furthermore, the rate at which electrons are moved within a metal or semiconductor is determined by the thickness of the Schottky barrier and the energy of the hot electrons [11].

6.2.3 Pure-Metal Plasmonic for Photocatalysis

Catalysis in any chemical reaction can also be supported by single-phase or pure-metal nanoparticles. The phenomena of excitation of LSPR can aid in the enhancement of catalytic reaction under the light. For this purpose, two mechanisms have been so far suggested and adopted (i) plasmonic heating and (ii) hot electron transfer.

Plasmonic heating is the process in which plasmonic nanocrystals of metal are converted through a photothermal process. The decay of LSPR is caused by the relaxation and thermalization of phonons in the lattice, as well as the relaxation of electrons. The temperature is set to be nearly 500 °C on a plasmonic metal surface to create resonance excitation in case of pulsed laser light. This high temperature leads to such a thermal reaction in which an ideal substrate can work [12].

Hot electron transfer is occurred when photoinduced electrons are transferred and is the same as plasmonic sensitization. When hot electrons are stimulated towards the lowest unoccupied molecular orbital (LUMO) of an adsorbate, they form transient anions. In this way, the chemical bond gets weakened; thus, the transient anions of lower energy are formed. Furthermore, as a result of the transient anions decay, and electrons injected into metal nanocrystals transfer energy in the molecular vibrational modes [6].

6.3 Background of Plasmonic Photocatalysts

Year	Scientists	Progress	References
1971	Endriz and Spicer	Photo-emission peak localized at the energies close the surface intensity of peak depends on the friction of surface	[13]
1990	Kostecki and Augustynski	Plasmon-induced photoelectrochemical activity of the rough Ag film electrode Photocurrent reduction of CO_2 was maximized	[14]
2004	Tian and Tatsuma	Efficiency of Au, Ag nanoparticles by incident from photon to current	[15]
2005	Schaadt	Hot electron injection by Au nanoparticles to TiO_2 Optical absorption by localized surface plasmon resonance	[16]
2006	Derkacs	Better behaviour of Si solar cells by plasmon-induced light by the metal nanoparticles	[17]
2008	Awazu	Plasmonic photocatalysis was introduced Oxidation reaction of methanol and formaldehyde was reported by Zhu's group	[18]
2010	Garcia's group	From water generation of solar hydrogen	[19]
2011	Brongersma's group	Photoelectrochemical cell (PEC) by localized surface plasmon resonance	[20]
2012	Wu, Cushing and coworker	Mechanism of plasmonic resonance transformation of energy	[21]

(continued)

(continued)

Year	Scientists	Progress	References
2012	Moskovits's group	Evaluation of oxygen and hydrogen at same time	[22]
2013	Wu, Cushing, and coworkers	Improvement in solar water splitting in PEC	[22]
2015	Wu, Cushing, and coworkers	Excellent efficiency of solar energy conversion	[23]

6.4 Applications of Plasmonic Photocatalysis

The plasmonic metal behaves as photosensitizer in the semiconductor photo-catalyst and photo-electrochemical cells. These photosensitizers can focus and set up the absorption rate of light and are widely used in photocatalyst and photo-electrochemical cells.

6.4.1 Solar Water Splitting

Most famous plasmonic metals are Au, Ag, Cu, and Al; they don't have the capability for the reduction and oxidation of water. The direct oxidation of water is complicated due to the limited life time of plasmonic hot holes, which is why all of these plasmonic metals, including Au, Ag, Cu, and Al, cannot be employed for photocatalytic water splitting. They are combined with other semiconductor and formed heterostructure for the photocatalytic process of solar splitting of water. For the photosensitizer of the semiconductor having large band gap, gold nanostructures are the best choice. It can enhance the rate of absorption of light of semiconductors such as ZnO and TiO_2 to the range of visible region as well as infrared region (almost 1100 nm). The cobalt photocatalyst can detect oxygen accumulated on the surface of Au nanorods. On the end of nanorods, Pt-designed TiO_2 is placed. The gap in between the oxygen evaluation and TiO_2 endorsed the partition of electrons and holes. Plasmonic hot electrons take part in the water reduction when they are poured to the TiO_2 while holes are poured into the catalyst, i.e. for oxygen evaluation which leads to the water reduction. Under solar sunlight, i.e. AM 1.5G, each nanorod produces around 51,013 hydrogen molecules autonomously in a unit centimetre square of space. The LSPR peak for Au nanoparticles is frequently larger than or equal to 520 nm; whereas, the band edge potential for TiO2 in anatase form is typically 386 nm. [24]. The heterojunction of Au-undoped TiO_2 has a plasmonic transfer of energy due to hot carrier insertion. Only one of both light trapping and PIRET phenomena can occur in Au-undoped TiO_2 heterojunction. This is because of the fact that there is no such overlapping between LSPR of Au and the absorption spectrum of titanium dioxide. Thus, the limited enhanced region can be achieved to covert energy in the heterojunction of Au-undoped TiO_2. This can be resolved by replacing gold with silver. Because of its significantly higher LSPR, silver can lead to higher plasmonic enrichment. This is because the greater LSPR band of silver (Ag) overlaps with TiO_2, allowing PIRET and light trapping to enhance solar conversion for TiO_2. For this purpose, nanoparticles of silver are assembled on an array of TiO_2 anode to fabricate a plasmonic photoanode of Ag-TiO_2. According to research, efficiency can lead up to 3.3 times in case of Ag-TiO_2 rather than using single-phase array of TiO_2 [25].

Normally, ZnO and TiO_2 have a broader band gap as compared to $BiVO_4$, Fe_2O_3, and WO_3. Using all three plasmonics, different flexible plasmonic photoelectrodes and photocatalysts may be constructed, and energy can be transported. For example, Fe_2O_3 is an n-type material and has a relatively smaller diffusion length for minor charges. In order to create a photoelectrode, Fe_2O_3 thin film is synthesized. Thus, light conversion can be assisted by the phenomena of PIRET and light trapping. After this incorporation of Au nanoparticles in ferric oxide, the excessive enhancement for plasmonic can be done by PIRET. However, the phenomenon of light trapping may be achieved on a massive scale by utilizing gold nanorods and nanospheres. A 300 nm high and 700 nm pitched nanopillar of gold array was made on a substrate coated with Fe_2O_3 film. The difference was compared between reversible hydrogen electrode of planar Fe_2O_3 film and plasmonic Au-Fe_2O_3 photoelectrode at 1.5 V, and the results were 40% enhanced in the latter [26].

Later research has shown the efficiency was 200% higher in Au-coated polyoxometalate-WO_3 photoanode when compared to photoanode based on bare WO_3. All these results show a better capability of plasmonic photoanode. The effective Fermi level is equilibrated under constant light, which tends to raise Fermi level, resulting in an increase in electron induction towards redox molecules. In this way, the electron hole recombination rate is reduced. Thus, in the state of steady illumination, electrons can be assembled on both metals and semiconductors. Electrons that are huge in number can be transferred to metal nanoparticles and then can be quickly induced in the solvent. This reduces the accumulation of electrons on the surface of the semiconductor, which reduces electron hole recombination. This effect of Fermi-level equilibration can be applied to both water splitting as well as redox reactions [27].

6.4.2 Environmental Applications

Generally, four types of contaminants are present in the air, such as bacteria, fungi, and viruses (pathogens), hydrocarbon compounds, in-organic anions, and heavy metals. There are different methods such as distillation, chemical oxidation, filtration, oxidation method, and biotechnology. All of these treatments are harmful and expensive; thus, a method for removing poisonous dyes was needed. Photocatalysis drag attention is due to its unique properties of removal of dyes. It is cost-effective and efficient technique for the removal of dyes by using the sun light. Pathogens, such as bacteria, fungi, and viruses, are inactivated, converting nitrate and nitrite in nitrogen, as well as contaminated heavy metals, to non-polluted metals. According to the thermodynamics, the electrochemical potential should be much less than 1.23 V. The theoretical efficiency of the photocatalytic process is significantly greater. It provides a number of varieties of photocatalytic semiconductors that are more flexible, and it attains less energy levels. Plasmonic metals can be utilized as catalytic surface light antenna during the photocatalysis process. When the reduction of pollutants is happened due to plasmonic photocatalysis, there are three main ways for this decomposition. Which are discussed as following: When plasmonic carriers are emitted from the surface of plasmonic semiconductor, can be helpful in the decomposition of organic pollutants. It can be removed by transferring the hot carriers from the metal surface to absorb the adsorbed surface oxygen to trap hot electrons to superoxide radical. Chen et al. reported the fabrication of Au nanoparticles reinforced respectively on the SiO_2 and ZrO_2. HCHO may be decomposed into CO2 by the excited LSPR of Au nanoparticles. All three mechanisms cannot take place during this decomposition when SiO_2 is used as insulating material. HCHO can be decomposed on the surface of Au by using plasmonic hot carriers. Mesoporous Au nanoparticles have been exposed to sunlight. The holes and superoxide radicals created are the product of an excited LSPR [28].

6.4.3 CO₂ Reduction Under Sunlight

Carbon dioxide can be converted into different forms of fuels through a number of techniques, including biocatalytic conversion, heterocatalytic reduction, electrocatalytic reduction, and photocatalytic reduction. Heterocatalysts, such as $Cu/ZnO/ZrO2$, are used to convert carbon dioxide into fuel, and they typically need a high temperature of approximately 350 °C. On the other hand, biocatalyst requires high cost and expensive procedure to convert CO_2 with enzymes. Moreover, its process is very time consuming, and the real issue can be enzyme stability for longer use. The electrochemical conversion of CO_2 into fuel, on the other hand, needs more energy, resulting in a less efficient current and thermodynamic reaction. An uphill reaction is required for reduction of CO_2 into liquid fuels. Thus, it requires high temperature and huge energy. On the contrary to this, the above stated uphill reaction can easily be manipulated by the use of photocatalysis [29].

At a neutral pH, CO_2 can be reduced by a single electron with an electrochemical potential of -1.9 eV versus normal hydrogen electrode (NHE). However, there is no single semiconductor to fulfil these requirements on a commercial scale as they do not have specific required band energies. As a result, a photocatalytic reduction of CO_2 occurs in which multiple electrons are processed with the assistance of protons. Water serves as an alternative source for protons. In this approach, CO_2 can be converted into hydrocarbons while water serves as a reducing agent. However, there is still the possibility of hydrogen synthesis from water as a competitor to CO_2-derived hydrocarbons. This is because the semiconductor used for reduction of CO_2 also has the ability to reduce water into hydrogen as the electrochemical potential for CO_2 reduction is more negative when compared with that of water [30].

Thus, the basic need is to lessen the production of hydrogen after controlling different parameters with the choice of perfect catalyst. In a heterogeneous catalysis reaction, plasmonic materials such as Au, Cu, Pt, and Ag can be used to solve this problem. Plasmonic materials help in the reaction kinetics, selectivity, and pathways which can lead to optimum photoconversion. Carbon dioxide can be reduced into different hydrocarbon compounds by adopting different electrochemical potential. Experimentally, TiO_2 and Au by themselves were unable to reduce CO_2, but Au-loaded TiO_2 was able to do so. Thorough plasmonic Au-TiO_2 heterojunction, CO_2 was converted into CH_4 under 532 nm illumination. Moreover, the efficiency of a single TiO_2 can be enhanced by 24-fold in case of Au-TiO_2, which proves the ultimate benefits of plasmonic in CO_2 conversion [31].

6.5 Conclusion

Energy requirement and environmental pollution are main problems and in order to deal with these problems a futuristic approach is needed to be operated. Photocatalysis has served as the answer to all such problems in future. However, there can be some limitations in using the semiconductor photocatalysis. The incorporation of metals such as silver, gold, platinum, and copper is discussed in this chapter as a way to circumvent the limits of semiconductor catalysts. When injected in the semiconductor, all these plasmonic materials can efficiently result from PIRET, LSPR, and electron trapping phenomena.

Solar conversion, pollution treatment, and carbon dioxide reduction can all benefit from plasmonic materials. When there are less chances of recombination in the case of plasmonics, wide band gap semiconductors can have more activity. It can be seen through a number of researches that Au incorporated TiO_2 had 24-fold higher efficiency than planar TiO_2. This was because of the higher LSPR peaks in Au deposition on the surface of TiO_2. Because of its higher LSPR and stronger electron trapping characteristics, silver is also a superior plasmonic material. The reduction of carbon dioxide into fuels can be so tricky because of the chance of simultaneous reduction of water and CO_2. However, the answer to this issue can be given by introducing plasmonic materials, which can efficiently convert carbon dioxide to hydrocarbons.

References

1. Tang, X., Wang, Z., & Wang, Y. (2018). Visible active N-doped TiO_2/reduced graphene oxide for the degradation of tetracycline hydrochloride. *Chemical Physics Letters, 691*, 408–414.
2. Sudhaik, A., Raizada, P., Shandilya, P., Jeong, D.-Y., Lim, J.-H., & Singh, P. (2018). Review on fabrication of graphitic carbon nitride based efficient nanocomposites for photodegradation of aqueous phase organic pollutants. *Journal of Industrial and Engineering Chemistry, 67*, 28–51.
3. Abebe, B., Murthy, H. A., & Amare, E. (2020). Enhancing the photocatalytic efficiency of ZnO: Defects, heterojunction, and optimization. *Environmental Nanotechnology, Monitoring & Management*, 100336.
4. Kundu, J., & Pradhan, D. (2013). Influence of precursor concentration, surfactant and temperature on the hydrothermal synthesis of CuS: Structural, thermal and optical properties. *New Journal of Chemistry, 37*(5), 1470–1478.
5. Li, J., & Wu, N. (2015). Semiconductor-based photocatalysts and photoelectrochemical cells for solar fuel generation: A review. *Catalysis Science & Technology, 5*(3), 1360–1384.
6. Xiao, M., Jiang, R., Wang, F., Fang, C., Wang, J., & Jimmy, C. Y. (2013). Plasmon-enhanced chemical reactions. *Journal of Materials Chemistry A, 1*(19), 5790–5805.
7. Zhang, N., Han, C., Fu, X., & Xu, Y. J. (2018). Function-oriented engineering of metal-based nanohybrids for photoredox catalysis: Exerting plasmonic effect and beyond. *Chem, 4*(8), 1832–1861.
8. Amiri, O., Salavati-Niasari, M., Mir, N., Beshkar, F., Saadat, M., & Ansari, F. (2018). Plasmonic enhancement of dye-sensitized solar cells by using Au-decorated Ag dendrites as a morphology-engineered. *Renewable Energy, 125*, 590–598.
9. Navalon, S., de Miguel, M., Martin, R., Alvaro, M., & Garcia, H. (2011). Enhancement of the catalytic activity of supported gold nanoparticles for the Fenton reaction by Light. *Journal of the American Chemical Society, 133*(7), 2218–2226.
10. Neaţu, S., Maciá-Agulló, J. A., Concepción, P., & Garcia, H. (2014). Gold–copper nanoalloys supported on TiO_2 as photocatalysts for CO2 reduction by water. *Journal of the American Chemical Society, 136*(45), 15969–15976.
11. Wang, X., & Caruso, R. A. (2011). Enhancing photocatalytic activity of titania materials by using porous structures and the addition of gold nanoparticles. *Journal of Materials Chemistry, 21*(1), 20–28.
12. Wu, X., Redmond, P. L., Liu, H., Chen, Y., Steigerwald, M., & Brus, L. (2008). Photovoltage mechanism for room light conversion of citrate stabilized silver nanocrystal seeds to large nanoprisms. *Journal of the American Chemical Society, 130*(29), 9500–9506.
13. Endriz, J., & Spicer, W. (1971). *Physical Review B, 4*, 4159–4184.
14. Kostecki, R., & Augustynski, J. (1995). *Journal of Applied Physics, 77*, 4701–4705.
15. Tian, Y., & Tatsuma, T.(2004). *Chemical Communications*, 1810–1811.
16. Tian, Y., & Tatsuma, T. (2005). *Journal of the American Chemical Society, 127*, 7632–7637.
17. Derkacs, D., Lim, S., Matheu, P., Mar, W., & Yu, E. (2006). *Applied Physics Letters, 89*, 093103.
18. Awazu, K., Fujimaki, M., Rockstuhl, C., Tominaga, J., Murakami, H., Ohki, Y., Yoshida, N., & Watanabe, T. (2008). *Journal of the American Chemical Society, 130*, 1676–1680.
19. Gomes Silva, C.U., Juárez, R., Marino, T., Molinari, R., & García, H. (2010). *Journal of the American Chemical Society, 133*, 595–602.
20. Thomann, I., Pinaud, B. A., Chen, Z., Clemens, B. M., Jaramillo, T. F., & Brongersma, M. L. (2011). *Nano Letters, 11*, 3440–3446.
21. Cushing, S. K., Li, J., Meng, F., Senty, T. R., Suri, S., Zhi, M., Li, M., Bristow, A. D., & Wu, N. (2012). *Journal of the American Chemical Society, 134*, 15033–15041.
22. Govorov, A.O., Zhang, H., & Gun'ko, Y. K. (2013). *Journal of Physical Chemistry C, 117*, 16616–16631.
23. Cushing, S. K., Bristow, A. D., & Wu, N. (2015). *Physical Chemistry Chemical Physics: PCCP, 17*, 30013–30022.
24. Mubeen, S., Lee, J., Singh, N., Krämer, S., Stucky, G. D., & Moskovits, M. (2013). *Nature Nanotechnology, 8*, 247–251.

25. Solarska, R., Bienkowski, K., Zoladek, S., Majcher, A., Stefaniuk, T., Kulesza, P. J., & Augustynski, J. (2014). *Angewandte Chemie, International Edition, 53*, 14196–14200.
26. Liu, G., Li, P., Zhao, G., Wang, X., Kong, J., Liu, H., Zhang, H., Chang, K., Meng, X., & Kako, T. (2016). *Journal of the American Chemical Society, 138*, 9128–9136.
27. Subramanian, V., Wolf, E. E., & Kamat, P. V. (2004). *Journal of the American Chemical Society, 126*, 4943–4950.
28. Hu, C., Peng, T., Hu, X., Nie, Y., Zhou, X., Qu, J., & He, H. (2009). *Journal of the American Chemical Society, 132*, 857–862.
29. Habisreutinger, S. N., Schmidt'Mende, L., & Stolarczyk, J. K. (2013). *Angewandte Chemie, International Edition, 52*, 7372–7408.
30. Neatu, S., Maciá-Agulló, J. A., Concepción, P., & Garcia, H. (2014). *Journal of the American Chemical Society, 136*, 15969–15976.
31. Long, R., Li, Y., Liu, Y., Chen, S., Zheng, X., Gao, C., He, C., Chen, N., Qi, Z., Song, L., Jiang, J., Zhu, J., & Xiong, Y. (2017). *Journal of the American Chemical Society, 139*, 4486–4492.

Chapter 7
Challenges and Future Prospects

Science has always been curious about nature, and the answer to the paradox that material physics can make earth a better place for human survival. Since its inception, nanotechnology has grabbed interest as a viable solution to any problem. Miniaturing of different devices has solved many issues like that. Nanomaterials can be classified in different types according to their size. It can be three-dimensional (3D) bulk, two-dimensional (2D) nanosheets, one-dimensional (1D) nanorods, and zero-dimensional (0D) quantum dots. All these materials have their own physical, chemical, and structural properties.

Different methods have been adopted over time to synthesize nanomaterials. It may include physical, chemical, biological, and electrochemical methods. However, physical processes are adopted for better yield. Physical production methods of nanoparticles include gamma radiations, laser ablation, chemical vapour deposition, and condensation.

Nanotechnology can be widely used in agriculture for a better yield of crops. It may be used in medicine to treat cancer, and nanobots can be introduced into the human body to treat other diseases. Electronic devices have become portable, flexible, and wearable gadgets with the help of nanoscience and material physics. It can also help in fabricating efficient energy devices to produce, store, and convert solar energy. However, industrialization has increased pollution over the years on the whole planet. Issues of energy, polluted water, and environmental pollution need to be answered immediately for future survival. Photocatalysis has shown to be a boon in answering all of above-mentioned issues. It is an environmentally friendly economic technique that utilizes sunlight to excite semiconductors to treat water, reduce carbon dioxide, degrade pollutants, and harvest energy. Some factors can affect the performance of any catalyst. The band gap, catalyst pH, reaction temperature, chosen technique, doping, and composite materials are only a few examples. However, there are several drawbacks to using this strategy such as light absorption, recombination rate, mobility, adsorption, etc. All these issues can be fixed using

M. B. Tahir et al., *New Insights in Photocatalysis for Environmental Applications*, SpringerBriefs in Applied Sciences and Technology, https://doi.org/10.1007/978-981-19-2116-2_7

different methods including introduction of plasmonic, elemental doping, composites, and heterojunction. Different semiconductors have been used over the years to resolve the above-discussed problems. The most famous semiconductor materials are metal oxides and metal sulphides. However, nitrides have also been used in the past few years. It has been seen that metal oxides have comparatively larger band gaps than metal sulphides. That is the reason metal oxides are irradiated under ultraviolet radiations. Metal sulphides with narrower band gaps, on the other hand, are bombarded with visible light. Some studies have also revealed that metal oxides are more stable than metal sulphides. Moreover, sulphides cannot directly split water because of their greater absorption ability. Metal sulphides are widely used in energy storage applications and in supercapacitors. Carbon-based materials are also used for the treatment of contaminated water. Fullerene and graphene have shown excellent catalytic performance to resolve many environmental issues.

Plasmonic metal loading on semiconductors is seen to be the most promising way for getting over catalyst limitations. The photocatalytic characteristics of this hybrid semiconductor–metal structure have been improved thanks to higher localized surface plasmonic resonance (LSPR), electron trapping, plasmon-induced resonance energy trapping (PIRET), and injection of hot electrons in the case of plasmonic metal. Electron trapping can make recombination slow; thus, photocatalytic activity may enhance. Moreover, LSPR overlapping with semiconductor can improve the absorption of light. Gold, silver, and platinum are the most often employed plasmonic metals, and silver has the most promising LSPR peaks for improving the absorption of light. Plasmonic-based photocatalysts are widely used in a few photocatalytic applications including CO_2 reduction and other environmental applications.

Overall, photocatalysis has served as a promising technique to answer several environmental crises globally. However, fabricating a material with all of the qualities of a perfect photocatalyst remains a challenge. The stability, cost, band gap, charge carrier mobility, maximum light absorption capability, and recombination rate are some of the issues that need to be resolved for any photocatalyst to be used at a large scale. Therefore, scientists need to fabricate such catalysts that require the least resources and energy which may lead to the better efficiency of catalysts.

Printed in the United States
by Baker & Taylor Publisher Services